Lecture Notes in Mathematics

A collection of informal reports and seminars
Edited by A. Dold, Heidelberg and B. Eckmann, Zürich

Series: Scuola Normale Superiore, Pisa
Adviser: E. Vesentini

T0233365

189

André Weil

Dirichlet Series and Automorphic Forms

Lezioni Fermiane

Springer-Verlag
Berlin · Heidelberg · New York 1971

AMS Subject Classifications (1970): 10D20, 12A85

ISBN 3-540-05382-4 Springer-Verlag Berlin · Heidelberg · New York
ISBN 0-387-05382-4 Springer-Verlag New York · Heidelberg · Berlin

Offsetdruck: Julius Beltz, Weinheim/Bergstr.

AVERTISSEMENT

Invité par la Scuola Normale Superiore de Pise à y faire un cours au printemps de 1969 dans le cadre des Lezioni Fermiane, j'y exposai la théorie dont j'avais déjà brièvement indiqué les résultats à Bombay au colloque de janvier 1968 (Bombay Colloquium on Algebraic Geometry, T.I.F.R., Bombay 1968, pp. 409-426). Le but final en était d'étendre à tous les corps de nombres et de fonctions ("A-corps" ou "corps globaux") les résultats classiques de Hecke sur les transformées de Mellin des séries de Dirichlet, et plus particulièrement le complément que j'y avais apporté dans une note de 1967 (Math. Ann. 168, pp. 149-156).

A Pise, il avait été question de publier ce cours tel qu'il avait été fait, en italien. Mais j'en repris le sujet à Princeton en 1969-70, et une rédaction provisoire en fut alors préparée par T. Miyake et H. Naganuma. C'est en substance cette rédaction qu'on trouvera ici, quelque peu remaniée et complétée par l'adjonction des Chapitres VIII et XI. Il n'est que juste néanmoins qu'elle paraisse dans le cadre des Lezioni Fermiane, puisque c'est à ce titre que j'eus l'occasion de traiter ce sujet avec quelque détail pour la première fois. Je suis heureux d'adresser ici mes remerciments chaleureux, avant tout à mes collègues de Pise, et tout particulièrement à E. Vesentini, pour leur invitation et la cordialité de leur accueil, et également à T. Miyake et H. Naganuma pour leur collaboration et le soin intelligent apporté à la rédaction de leurs notes.

Comme je l'indiquais déjà dans ma conférence de Bombay, les recherches exposées ici ne font guère, à bien des égards, que doubler une partie de celles de H. Jacquet et R. Langlands, qui forment le contenu de leur monumental ouvrage Automorphic Forms on GL(2), 2 lb. 3 oz., Lecture-Notes No. 114, Springer 1970. Aussi jugera-t-on peut-être que leur publication rendait celle-ci superflue, et cela d'autant plus que leur point de vue (celui de la théorie des représentations) va sans doute plus au fond des choses que le mien, qui est essentiellement élémentaire. Mais c'est justement en raison de cette différence de points de vue, dans un domaine où le dernier mot ne sera sûrement pas prononcé d'ici longtemps, que je n'ai pas cru tout à fait inutile de mettre mes démonstrations à la disposition du public. Quant aux "priorités" (s'il est quelqu'un qui s'y intéresse), il suffira de dire que Jacquet et Langlands m'ont constamment tenu au courant de leur travail, et qu'en plus d'emprunts purs et simples (par exemple tout ce qui concerne l'"équation fonctionnelle locale" pour GL(2, $\mathbb{R}$) et GL(2, $\mathbb{C}$)), je n'ai pu manquer, consciemment ou non, d'en subir l'influence en mainte occasion. Faute d'être en état de rien dire de plus précis, je dois me contenter de leur exprimer ma reconnaissance, et de renvoyer à leur ouvrage (qu'on complètera utilement par R. Godement, Notes on Jacquet-Langlands, I. A. S. 1970) le lecteur désireux de comparer leurs résultats et les miens.

Enfin, c'est un agréable devoir pour moi de remercier Miss Margaret Murray à qui est due la dactylographie du présent volume et qui s'est acquittée de cette tâche parfois épineuse avec sa conscience coutumière.

Princeton, le 26 novembre 1970.

TABLE OF CONTENTS

CHAPTER I

THE CLASSICAL CASE

1. Let G be a semisimple group, K its maximal compact subgroup, so that G/K is the associated Riemannian symmetric space. In the classical theory of automorphic functions, one studies the functions on G/K with a prescribed local behavior (e.g. holomorphic functions if G/K is Hermitian and is given its "natural" complex structure), and a prescribed behavior with respect to some suitable discrete subgroup Γ. A typical case is given by the classical modular functions and modular forms; there, G can be taken to be $SL(2, \mathbb{R})$; G/K is the Poincaré half-plane; one may take for Γ the full modular group $SL(2, \mathbb{Z})$, or a congruence subgroup of that group. A rather full but somewhat longwinded function-theoretic treatment of that case was given in 1890-92 by Fricke and Klein; the arithmetical aspects, which are intimately tied up with the theory of complex multiplication, were considered by H. Weber in his Algebra, vol. III (for a modern treatment, cf. a forthcoming book by G. Shimura). The relation between modular forms and Dirichlet series with functional equations was discovered by Hecke, whose epoch-making work during the years 1930-1940, based on that discovery and that of the "Hecke operators," brought out completely new aspects of a theory which many mathematicians would have regarded as a closed chapter long before.

Even in the earlier days of the theory, attempts were made to extend it to suitable types of automorphic functions of more than one (complex) variable; an attempt by Hecke, in his early work, to extend correspondingly the theory of complex multiplication was only partly successful, for reasons which were not understood until much later. Further inroads into the extension of the classical theory to other groups than $SL(2, \mathbb{R})$ had to depend upon a deeper understanding of the semisimple groups and of their classification; the most decisive steps were taken by C. L. Siegel, in his famous work on the symplectic group, and by H. Maass, who was the first to take up systematically the study of non-holomorphic automorphic functions.

The role played by $SL(2, \mathbf{Z})$ in the theory of modular functions had already suggested to Hilbert the idea of substituting for $\mathbf{Z}$ the ring of integers in a more general algebraic number-field; as long as holomorphic functions were the only ones to be studied, this field had to be taken totally real, since otherwise the corresponding Riemannian symmetric space has no complex structure. Even in that case, it was soon noticed that serious technical difficulties arose when the number of ideal-classes is greater than one; when the attempt was made to extend the work of Hecke to such fields, it became clear that the language and notations of the classical theory had become inadequate. In a parallel case, that of Siegel's theory of quadratic forms, similar difficulties had already been removed by a systematic use of the adele concept, which makes it possible to treat all places of a number-field simultaneously without unduly emphasizing some of them (the infinite or "archimedean" ones); this is also the best way of dealing at the same time with number-fields and with function-fields, since the latter have no infinite places. Naturally this does not mean that, in studying any specific problem, one ought not to revert sometimes to a more traditional treatment, where some places in finite number (including the infinite ones, but not necessarily restricted to those) are singled out for special consideration.

In these lectures, the adele point of view will be adopted systematically, even though, at one stage (cf. Chapter VIII) the infinite places will have to receive a separate treatment. The main result will be one which generalizes (not quite completely) a theorem on modular forms, proved in my note of 1967 (loc. cit.) in continuation of Hecke's work. As the latter, naturally, was stated in terms of the classical theory, the connection between them cannot be perceived unless the means are given for shifting from one point of view to the other; this will be done now.

2. For technical reasons, we take as our starting point, not $SL(2)$, but the group $G = GL(2)$, considered as an algebraic group over the groundfield $k = \mathbf{Q}$. With the notations which are now customary in the theory of algebraic groups, G_k is then $GL(2, \mathbf{Q})$, and, if v is any place of $k = \mathbf{Q}$, G_v denotes the group $GL(2, k_v)$; in particular, we have $G_\infty = GL(2, \mathbf{R})$. For each natural prime p, put $K_p = GL(2, \mathbf{Z}_p)$; this is the maximal compact subgroup of the group $G_p = GL(2, \mathbf{Q}_p)$, consisting of the matrices $\begin{pmatrix} a & b \\ c & d \end{pmatrix}$ with a, b, c, d in $\mathbf{Z}_p$ (the ring of p-adic integers) and $ad - bc$ in $\mathbf{Z}_p^\times$ (the group of p-adic units). The "adelized" group G_A is then the union (more precisely, the "inductive limit") of the groups

$$\prod_{v \in P} G_v \times \prod_{p \notin P} K_p ,$$

when one takes for P all the finite sets of places of $\mathbf{Q}$ containing ∞. Then G_k is a discrete subgroup of G_A.

From the adele point of view, the theory of automorphic functions, or forms, on $GL(2)$ over $\mathbf{Q}$ is the theory of various types of functions on $G_k \backslash G_A$, i.e. functions on G_A which are left-invariant under G_k. In a loose sense, we may speak of the "space" of such functions (this can be narrowed down in various ways, e.g. to $L^2(G_k \backslash G_A)$, but we prefer not to do it here). In a loose sense again, this "space" may be "decomposed" according to the characters, or rather to the quasicharacters, of the center. Write $\mathfrak{z}$ for the center of G as an algebraic group; this is isomorphic to $GL(1)$ and consists of the scalar multiples $z.1_2$ of the unit-matrix $1_2 = \begin{pmatrix} 1 & 0 \\ 0 & 1 \end{pmatrix}$. Then the center $\mathfrak{z}_A$ of G_A consists of the elements $\mathfrak{z} = z.1_2$, where z is an idele of $\mathbf{Q}$; it may thus be identified, whenever convenient, with the idele group $\mathbf{Q}_A^\times$ of $\mathbf{Q}$; the center $\mathfrak{z}_k$ of G_k may similarly be identified with $k^\times = \mathbf{Q}^\times$. Let α be a quasicharacter of $\mathfrak{z}_A$

(i.e. a morphism of that group into $\mathbb{C}^\times$); if a function φ on $G_k \backslash G_A$ satisfies the condition

$$\varphi(g\mathfrak{z}) = \varphi(g)\alpha(\mathfrak{z}) \tag{1}$$

for all $g \in G_A$ and $\mathfrak{z} \in \mathfrak{Z}_A$, α must clearly be trivial (i.e. constant and equal to 1) on $\mathfrak{Z}_k$. Thus the "space" of functions on $G_k \backslash G_A$ may be decomposed according to the quasicharacters of the idele-class group $k_A^\times / k^\times$; to each such quasicharacter α, we attach the functions φ on $G_k \backslash G_A$ satisfying (1). In particular, for $\alpha = 1$, we get the functions on $G_k \backslash G_A$ which are invariant under $\mathfrak{Z}_A$; as $\mathfrak{Z}_A$ is central, we need not specify here whether they are right-invariant or left-invariant. As $G/\mathfrak{Z}$ is the same as the projective group $G' = PGL(2)$, such functions could also be regarded as the functions on $G'_k \backslash G'_A$. In §§3-4, we will consider only such functions.

3. Call G_∞^o the subgroup of G_∞ given by $\det(g_\infty) > 0$, i.e. the connected component of 1_2 in G_∞; call Ω the open subgroup $G_\infty^o \times \prod K_p$ of G_A. Then we have $G_A = G_k \Omega$. In fact, $g \longrightarrow \det(g)$ is a morphism of G_A onto $\mathbb{Q}_A^\times$, and $\mathbb{Q}_A^\times$ is the direct product of $\mathbb{Q}^\times$ and of the subgroup $\mathbb{R}_+^\times \times \prod \mathbb{Z}_p^\times$ of $\mathbb{Q}_A^\times$. For any $g \in G_A$, write $\det g = r.u$, with $r \in \mathbb{Q}^\times$, $u_\infty > 0$, $u_p \in \mathbb{Z}_p^\times$ for all p; put

$$g' = \begin{pmatrix} r^{-1} & 0 \\ 0 & 1 \end{pmatrix} . g . \begin{pmatrix} u^{-1} & 0 \\ 0 & 1 \end{pmatrix} ;$$

then $\det g' = 1$. In view of the well-known "approximation theorem" for $SL(2, \mathbb{Q})$, we can find $\gamma_1 \in SL(2, \mathbb{Q})$ such that, for all p, the p-component of $\gamma_1^{-1} g'$ is in K_p. Put $\gamma = \begin{pmatrix} r & 0 \\ 0 & 1 \end{pmatrix} . \gamma_1$; then γ is in G_k and $\gamma^{-1} g$ in Ω, which proves our assertion.

Consequently, a function φ on $G_k \backslash G_A$ is uniquely determined by its values on Ω. Conversely, a function on Ω can be extended to a function on G_A, left-invariant under G_k, if and only if it is left-invariant under the group $G_k \cap \Omega$, which is no other than the "classical" modular group $SL(2, \mathbf{Z})$. In the same manner, we see that φ is invariant under the center of G_A if and only if the function it induces on Ω is invariant under the center of Ω.

Now we restrict our function-space again, by prescribing that φ shall be right-invariant under all the groups K_p, and prescribing also its behavior under the maximal compact subgroup of G_∞^o; this consists of the rotations

$$r(\theta) = \begin{pmatrix} \cos\theta & \sin\theta \\ -\sin\theta & \cos\theta \end{pmatrix} .$$

We do this by assuming

$$(2) \qquad \varphi(g.\, r(\theta)) = \varphi(g) e^{i\nu\theta}$$

for all $g \in G_A$ and all θ. As we have agreed, in §2, to take $\alpha = 1$, ν must be an even integer, since $r(\pi) = -1_2$.

As one sees at once, every element g_∞ of G_∞^o can be uniquely written in the form

$$(3) \qquad g_\infty = z.\begin{pmatrix} x & y \\ 0 & 1 \end{pmatrix}.\, r(\theta)$$

with $z > 0$, $x > 0$; in view of what we have found, this shows that φ, subject to the above conditions, is uniquely determined by the function $f(x, y)$ it induces on the subgroup B_∞^o of G_∞^o consisting of the matrices $\begin{pmatrix} x & y \\ 0 & 1 \end{pmatrix}$ with $x > 0$.

4. Let now a function $f(x, y)$ be given on B_∞^o; we wish to know whether it can be extended to a function φ on G_A, of the kind described in §§2-3. Take $g \in \Omega$, and write g_∞ in the form (3). Then we must have

$$\varphi(g) = f(x, y) e^{i\nu\theta} .$$

If σ is any element of $SL(2, \mathbf{Z})$, σg is also in Ω, so that we can write $(\sigma g)_\infty$ in the form

$$(\sigma g)_\infty = z' . \begin{pmatrix} x' & y' \\ 0 & 1 \end{pmatrix} . r(\theta') \ ;$$

then we must have $\varphi(\sigma g) = \varphi(g)$, and therefore

$$f(x', y') = f(x, y) e^{i\nu(\theta - \theta')} \ .$$

Put $\tau = y + ix$, $\tau' = y' + ix'$; these are points in the Poincaré half-plane, and it is easily seen that τ' is no other than the image of τ under σ; if $\sigma = \begin{pmatrix} a & b \\ c & d \end{pmatrix}$, this is

$$\tau' = \frac{a\tau + b}{c\tau + d} \ .$$

Write now

$$F(\tau) = x^{-\nu/2} f(x, y) \ ;$$

then the above relation between $f(x, y)$ and $f(x', y')$ can be rewritten as

$$F(\tau') = F(\tau)(c\tau + d)^\nu \ .$$

In other words, under the modular group, F must behave as a modular form of degree ν. Conversely, if this is assumed, the above calculations show that f can be extended uniquely to a function φ of the required kind.

In the classical theory, one also requires F to be holomorphic in the upper half-plane. One verifies easily that this amounts to prescribing a differential equation for φ as a function of g_∞, viz., $W\varphi = 0$ if W is the left-invariant operator on G_∞ defined by the element $\begin{pmatrix} 1 & -i \\ -i & -1 \end{pmatrix}$ of its complexified Lie algebra; in this equation, it is of course understood that all the components g_p of g at the finite places are kept constant. The further condition of the classical theory, pertaining to the behavior of $F(\tau)$ at the cusp $\tau = i\infty$ of the "fundamental domain" of the modular group could now be expressed as a bound for the order of magnitude of φ in a fundamental domain for G_k in G_A; such a domain is given by the main theorem of reduction theory, when this theorem is restated in terms of the adele group G_A; we refrain from giving more details here.

5. The above treatment can be generalized in various ways to congruence subgroups of $SL(2, \mathbf{Z})$; we will merely outline a mode of treatment of Hecke's group $\Gamma_0(N)$, which is of special relevance to what follows. For any integer $N > 1$, $\Gamma_0(N)$ is defined as consisting of the matrices $\begin{pmatrix} a & b \\ Nc & d \end{pmatrix}$, where a, b, c, d are integers, and $ad - Nbc = 1$. For each prime p dividing N, call K'_p the subgroup of K_p consisting of the matrices $\mathfrak{k} = \begin{pmatrix} u & v \\ Nw & t \end{pmatrix}$ in K_p, with u, v, w, t in $\mathbf{Z}_p$; if p^m is the highest power of p dividing N, this can also be described as consisting of the matrices $\begin{pmatrix} u & z \\ p^m w & t \end{pmatrix}$, with u, t in $\mathbf{Z}_p^\times$ and z, w in $\mathbf{Z}_p$. If p does not divide N, we put $K'_p = K_p$. Put now $\Omega' = G_\infty^o \times \prod K'_p$; then $G_A = G_k\Omega'$, just as before, and $G_k \cap \Omega' = \Gamma_0(N)$. Take a quasicharacter α of $\mathbf{Q}_A^\times/\mathbf{Q}^\times$, whose conductor divides N; on $\mathbf{Q}_\infty^\times$, let it be given by $\alpha_\infty(x) = (\text{sgn } x)^A . |x|^s$, with $A = 0$ or 1 and $s \in \mathbf{C}$ (it would be no real restriction, in this case, to assume $s = 0$). We now prescribe, for a function φ on $G_k\backslash G_A$, the following behavior: (a) it shall satisfy (1) for all $\mathfrak{z} \in \mathfrak{Z}_A$; (b) for p not dividing N, it shall be right-invariant under K'_p, i.e. under K_p; (c) for p dividing N, p^m as above, and $\mathfrak{k} = \begin{pmatrix} u & v \\ Nw & t \end{pmatrix}$ in K'_p, we prescribe $\varphi(g\mathfrak{k}) = \varphi(g)\alpha(t)$; (d) under the rotations $r(\theta)$, φ should behave according to (2), where ν is a given integer, which must be taken $\equiv A \bmod. 2$ since $r(\pi) = -1_2$ is in $\mathfrak{Z}_\infty$. Define $f(x, y)$ as before, and put

$$F(\tau) = x^{-(\nu+s)/2} f(x, y) .$$

Then a calculation, similar to the one in §4, gives

$$F(\tau') = F(\tau)(c\tau + d)^{\nu} \alpha(d)^{-1}$$

for all $\sigma = \begin{pmatrix} a & b \\ c & d \end{pmatrix}$ in the Hecke group $\Gamma_o(N)$; here $\alpha(d)$ denotes the character of the integers prime to N modulo N which is defined by

$$\alpha(d) = \prod_{p/N} \alpha_p(d) \ .$$

Conversely, if this is so, $f(x, y)$ can be uniquely extended to a function φ on $G_k \backslash G_A$, of the type described by conditions (a) to (d).

CHAPTER II

DIRICHLET SERIES

6. Once for all, we take as our groundfield an A-field k, i.e. an algebraic number-field (of finite degree over $\mathbb{Q}$) or an algebraic function-field of dimension 1 over a finite field of constants, according as its characteristic is 0 or >1. As usual, for each place v of k, k_v will denote the completion of k at v; if v is a finite place, we write r_v, $r_v^{\times}$, π_v for the ring of integers of k_v, the group of units in r_v, and a prime element of r_v, respectively; k_A and $k_A^{\times}$ are the adele ring and the idele group of k, respectively. To each finite place v we attach a "prime divisor" $\mathfrak{p}_v$; if $\mathfrak{p}$ is a prime divisor, we sometimes also write $\mathfrak{p}$ for the place to which it belongs. We denote by $\mathfrak{M}$ the group of divisors of k, i.e. the free abelian group generated by the prime divisors $\mathfrak{p}_v$, which we write multiplicatively. Consequently, we write $\mathfrak{m} \succ 1$ to denote that the divisor $\mathfrak{m}$ is positive, i.e. that the exponents of all $\mathfrak{p}_v$ in it are $\geqq 0$; we write $\mathfrak{M}_+$ for the semigroup of positive divisors. If $\mathrm{char}(k) = 0$, $\mathfrak{M}$ may be identified with the group of fractional ideals of k, and $\mathfrak{M}_+$ with the set of all ideals $\neq (0)$ in the ring of integers of k. For any divisor $\mathfrak{m}$, we put

$$|\mathfrak{m}| = N(\mathfrak{m})^{-1} \quad \text{if } \mathrm{char}(k) \text{ is } 0$$
$$= q^{-\deg(\mathfrak{m})} \quad \text{if k is a function-field over a field of constants with q elements.}$$

For each idele $x = (x_v)$, we define

$$\mathrm{div}(x) = \prod \mathfrak{p}_v^{\mathrm{ord}(x_v)} ,$$

the product being taken over all finite places. We write x_∞ for the idele x' given by $x'_w = x_w$ for all infinite places w, and $x'_v = 1$ for all finite places v; this is the projection of x onto the subgroup $k_\infty^\times = \prod_w k_w^\times$. Then the "idele module" $|x|_A$ (also written $|x|$, if this causes no confusion) is given by $|x|_A = |x_\infty| . |\mathrm{div}\, x|$. To simplify notations, we will usually write $|x|_v$ instead of $|x_v|_v$, if x is any idele or adele.

If $\mathfrak{m} \longrightarrow c(\mathfrak{m})$ is any mapping of $\mathfrak{M}_+$ into $\mathbb{C}$, the formal Dirichlet series

$$\sum_{\mathfrak{m} \in \mathfrak{M}_+} c(\mathfrak{m})\, |\mathfrak{m}|^s$$

will be called a <u>Dirichlet series belonging to</u> k provided it is absolutely convergent for some s; this will be so if and only if $|c(\mathfrak{m})| \leqq C\, |\mathfrak{m}|^{-a}$ for some $C > 0$ and some $a \in \mathbf{R}$.

Let $\mathfrak{p}$ be a prime divisor. If $c(\mathfrak{m}'\mathfrak{p}^n) = c(\mathfrak{m}')c(\mathfrak{p}^n)$ for every $n \geqq 0$ and every positive divisor $\mathfrak{m}'$ disjoint from $\mathfrak{p}$ (i.e. in which $\mathfrak{p}$ has the exponent 0), we have formally

$$\sum_{\mathfrak{m} \in \mathfrak{M}_+} c(\mathfrak{m})\, |\mathfrak{m}|^s = \sum_{\mathfrak{m}'} c(\mathfrak{m}')\, |\mathfrak{m}'|^s . \sum_{n \geqq 0} c(\mathfrak{p}^n)\, |\mathfrak{p}|^{ns} ,$$

where the first sum in the right-hand side is taken over the divisors $\mathfrak{m}'$ disjoint from $\mathfrak{p}$. When that is so, we say that the given Dirichlet series is weakly eulerian at $\mathfrak{p}$, with the Euler factor

$$\sum_{n \geqq 0} c(\mathfrak{p}^n)\, |\mathfrak{p}|^{ns} .$$

If at the same time this factor is (formally) equal to

$$(1 + a_1 |\mathfrak{p}|^s + \ldots + a_d |\mathfrak{p}|^{ds})^{-1} ,$$

with some constants $a_1, \ldots, a_d$, we say that the Dirichlet series is eulerian at $\mathfrak{p}$ or that it has the Euler property of degree d at $\mathfrak{p}$. In these lectures, only the case $d \leqq 2$ will occur.

7. Once for all, we choose a non-trivial additive character ψ of k_A, trivial on k. Then every character ψ' with the same property can be written as $\psi'(x) = \psi(\xi x)$ with $\xi \in k^\times$. For each place v, we write ψ_v for the character of k_v induced by ψ on k_v. For v finite, we say that ψ_v is of order δ, or that ψ is of order δ at v, if ψ_v is trivial on $\pi_v^{-\delta} r_v$ but not on $\pi_v^{-\delta-1} r_v$; δ is 0 for almost all v.

Once for all, we will assume ψ chosen so that we have

$$\psi_w(x) = e^{-2\pi i x} \quad \text{when } k_w = \mathbf{R} ,$$

$$\psi_w(x) = e^{-2\pi i(x+\bar{x})} \quad \text{when } k_w = \mathbf{C} .$$

These conditions determine ψ uniquely if k is a number-field; if it is a function-field, they are empty, and then there are infinitely many permissible choices for ψ. For each finite v, let $\delta(v)$ be the order of ψ at v. Let d be the idele given by $d_v = \pi_v^{\delta(v)}$ for v finite and $d_w = 1$ for w infinite; we call d <u>a differental idele</u> belonging to ψ; it depends upon the choice of the prime elements π_v, but $\mathrm{div}(d)$ does not; with our choice of ψ in the number-field case, $\mathrm{div}(d)$ is then "the different" of k in the usual sense.

8. By a quasicharacter of a group, we understand a morphism of that group into $\mathbf{C}^\times$. Let ω be a quasicharacter of the "idele-class group" $k_A^\times/k^\times$, or (what amounts to the same) a morphism of $k_A^\times$ into $\mathbf{C}^\times$, trivial on $k^\times$. As usual, we write ω_v for the quasicharacter of $k_v^\times$ induced on $k_v^\times$ by ω. For each place v, choose $f_v \in k_v^\times$ as follows: if v is infinite, or if v is finite and ω_v is trivial on $r_v^\times$, we take $f_v = 1$; otherwise we take $\mathrm{ord}(f_v) > 0$ and such that ω_v is trivial on the subgroup $1 + f_v r_v$ of $r_v^\times$ but not on $1 + \pi_v^{-1} f_v r_v$. This defines an idele $f = (f_v)$, whose divisor $\mathfrak{f} = \mathrm{div}(f)$ is uniquely determined by ω and is called the conductor of ω.

If v is a finite place, not occurring in $\mathfrak{f}$, $\omega(\pi_v)$ is independent of the choice of the prime element π_v and will also be denoted by $\omega(\mathfrak{p}_v)$ if $\mathfrak{p}_v$ is the prime divisor attached to v. The group generated by the $\mathfrak{p}_v$, for v finite and not occurring in $\mathfrak{f}$, is the group of the divisors of k disjoint from $\mathfrak{f}$; we can extend uniquely the mapping $\mathfrak{p}_v \longrightarrow \omega(\mathfrak{p}_v)$ to a quasi-character of that group, which we also denote by ω; then we define a mapping ω of the semigroup $\mathfrak{M}_+$ into $\mathbf{C}$ by prescribing that $\omega(\mathfrak{m})$ should be as just defined, if the positive divisor $\mathfrak{m}$ is disjoint from $\mathfrak{f}$, and otherwise that it should be 0. We may then consider the Dirichlet series

$$\sum_{\mathfrak{m} \in \mathfrak{M}_+} \omega(\mathfrak{m})\,|\mathfrak{m}|^s = \prod_{\mathfrak{p}} (1 - \omega(\mathfrak{p})\,|\mathfrak{p}|^s)^{-1} ,$$

where the product in the right-hand side is taken over all the prime divisors $\mathfrak{p}$ outside $\mathfrak{f}$; this is known as the L-series belonging to ω; the above identity shows that it has everywhere the Euler property (of degree 1 or 0).

For every $\alpha \in \mathbf{C}$, the mapping $x \longrightarrow |x|^\alpha$ is a quasicharacter of $k_A^\times/k^\times$, which we denote by ω_α; its conductor is 1, and we have then $\omega_\alpha(\mathfrak{m}) = |\mathfrak{m}|^\alpha$ for all divisors $\mathfrak{m}$. The L-series attached to ω_α is $Z(s + \alpha)$, where Z is the zeta-function of k. If ω is any quasicharacter of $k_A^\times/k^\times$, and $|\omega|$ is its (ordinary) absolute value (i.e. $(\omega\bar{\omega})^{1/2}$), we have $|\omega| = \omega_\sigma$, i.e. $|\omega(x)| = |x|_A^\sigma$, with some $\sigma \in \mathbf{R}$; this implies of course the absolute convergence of the L-series, and of the infinite product which expresses it, for $\mathrm{Re}(s) > 1 - \sigma$.

9. On the group Ω_k of the quasicharacters of $k_A^\times/k^\times$, we will define, not merely a topology, but a structure of complex variety of dimension 1; this is done as follows. The connected components of Ω_k will be the cosets of the subgroup consisting of the "elementary" quasi-characters ω_s. On this subgroup we define a complex structure by

prescribing that the morphism $s \longrightarrow \omega_s$ of $\mathbf{C}$ onto that group shall be complex-analytic, i.e. holomorphic; the kernel of that morphism is $\{0\}$ or $2\pi i(\log q)^{-1}\mathbf{Z}$ according as k is a number-field or a function-field over a field of constants with q elements; thus that group is isomorphic to $\mathbf{C}$ in the former case and to $\mathbf{C}^{\times}$ in the latter case. On the cosets of that group, we define the complex structure in the obvious manner, by translation, so that, for every ω, the mapping $s \longrightarrow \omega_s\omega$ is complex-analytic.

Now let a Dirichlet series, belonging to k, be given by its coefficients $c(\mathfrak{m})$; consider, for all $\omega \in \Omega_k$, the series

$$Z(\omega) = \sum_{\mathfrak{m} \in \mathfrak{M}_+} c(\mathfrak{m})\omega(\mathfrak{m}) .$$

As we have assumed that the Dirichlet series is convergent somewhere, we have $|c(\mathfrak{m})| \leqq C|\mathfrak{m}|^{-a}$ with some $C > 0$ and $a \in \mathbf{R}$. Putting now $|\omega| = \omega_\sigma$ with $\sigma \in \mathbf{R}$, it is clear that $Z(\omega)$ converges whenever $\sigma > a + 1$ and defines a holomorphic function on the part of Ω_k defined by that condition. It is easy to show that the latter function determines the coefficients $c(\mathfrak{m})$ uniquely (while the original Dirichlet series, considered as a function in part of the s-plane, does not). We will call $Z(\omega)$ the <u>extended</u> Dirichlet series (or sometimes, by abuse of language, merely the Dirichlet series) defined by the coefficients $c(\mathfrak{m})$.

Clearly, if the Dirichlet series with the coefficients $c(\mathfrak{m})$ is eulerian at $\mathfrak{y}$, with the Euler factor $(1 + \sum_1^d a_i|\mathfrak{y}|^{is})^{-1}$, the extended Dirichlet series can be written as

$$\sum_{\mathfrak{m}'} c(\mathfrak{m}')\omega(\mathfrak{m}').(1 + \sum_{i=1}^{d} a_i\omega(\mathfrak{y})^i)^{-1} ,$$

where $\omega(\mathfrak{y})$, as explained, is 0 if $\mathfrak{y}$ occurs in the conductor of ω, and the first sum is taken over the positive divisors disjoint from $\mathfrak{y}$.

10. It is known that, in many cases, an extended Dirichlet series $Z(\omega)$ can be continued analytically, as a meromorphic or even as a holomorphic function, over the whole of Ω_k, and satisfies a functional equation. This is so, in particular, when we take $c(\mathfrak{m}) = 1$ for all $\mathfrak{m} \in \mathfrak{M}_+$; then the original Dirichlet series is the zeta-function of k, and the extended Dirichlet series, restricted to the various connected components of the group Ω_k, defines there all the L-series attached to k. It is well-known that this extended Dirichlet series can be analytically continued to the whole of Ω_k as a meromorphic function with only two poles, at $\omega_o = 1$ and at ω_1, and that it satisfies a functional equation relating together the values $Z(\omega)$ and $Z(\omega_1\omega^{-1})$ (cf. e.g. my Basic Number Theory, Chap. VIII). For the convenience of the reader, we recall here one result which occurs in the proof of that functional equation, and which will be needed in Chapters VII and X. It concerns the integral

$$i(z) = \int_{r_v^\times} \omega_v(u)\psi_v(zu)d^\times u$$

where v is a finite place, $d^\times u$ is a Haar measure on $k_v^\times$, $z \in k_v$, and ψ, ω are as above; since $d^\times u$ coincides on $r_v^\times$ with an additive Haar measure on k_v, this may be regarded as the Fourier transform on k_v of the function equal to $\omega_v(u)$ on $r_v^\times$ and to 0 outside $r_v^\times$. Assume for instance that $d^\times u$ is so normalized that the measure of $r_v^\times$ is 1. Take d, f as before, so that $\mathrm{ord}(d_v)$ is the order of ψ_v, and $\mathrm{ord}(f_v)$ is 0 if $\omega_v = 1$ on $r_v^\times$, and otherwise the smallest ν such that $\omega_v = 1$ on $1 + \pi_v^\nu r_v$. Put $q_v = |\pi_v|^{-1}$. The results are then as follows:

a) if v does not occur in the conductor of ω, i.e. if $\mathrm{ord}(f_v) = 0$, $i(z)$ is 1, $-(q_v - 1)^{-1}$ or 0 according as $\mathrm{ord}(z)$ is $\geqq -\mathrm{ord}(d_v)$, $-\mathrm{ord}(d_v)-1$ or $< -\mathrm{ord}(d_v)-1$;

b) if $\mathrm{ord}(f_v) \geqq 1$, $i(z)$ is 0 if $\mathrm{ord}(z) \neq -\mathrm{ord}(d_v f_v)$;

c) if $\mathrm{ord}(f_v) \geqq 1$ and $\mathrm{ord}(z) = -\mathrm{ord}(d_v f_v)$, we have

$$i(z) = \kappa_v^{-1}(1 - q_v^{-1})^{-1}|f|_v^{-1/2}\omega(-d_v f_v z)^{-1} ,$$

where κ_v is a "Gaussian sum" (normalized so that it has the absolute value 1) depending upon ω and also (in an obvious manner) upon the choice of ψ, d and f.

For an infinite place w, one is led to put $\kappa_w = i^{-|A|}$, where A is 0 or 1 according as $\omega_w(-1)$ is 1 or -1 if $k_w = \mathbf{R}$, and A is the integer such that $x^A \omega_w(x)$ is 1 when $x\bar{x} = 1$ if $k_w = \mathbf{C}$. If we also put $\kappa_v = 1$ when v is a finite place outside the conductor of ω, the constant occurring in the functional equation for the L-series attached to ω is then $\varepsilon(\omega) = \kappa\omega(df)$, with $\kappa = \prod_v \kappa_v$, where the product is taken over all the places of k; it depends only upon ω, and not upon the choice of ψ, d and f.

CHAPTER III

BASIC CONCEPTS

11. Let $G = GL(2)$ be the general linear group in 2 variables, considered as an algebraic group over k. We write $\mathfrak{z}$ for its center, which consists of the elements $z.1_2$ and can thus be identified with the multiplicative group $G_m = GL(1)$. As usual we write G_A, $\mathfrak{z}_A$ for the corresponding adelized groups; $\mathfrak{z}_A$ can be identified with the idele group $k_A^\times$ of k. We have $G_k = GL(2, k)$, $G_v = GL(2, k_v)$; their centers are $\mathfrak{z}_k$, $\mathfrak{z}_v$, and may be identified with $k^\times$, $k_v^\times$, respectively. We write G_∞, $\mathfrak{z}_\infty$ for the products $\prod G_w$, $\prod \mathfrak{z}_w$, taken over the infinite places of k; the latter may be identified with $k_\infty^\times$; if k is a function-field, it should be understood that both of these groups are $\{1\}$.

Once for all, we choose an idele $a = (a_v)$ with $a_v = 1$ for almost all places including all the infinite ones, and $\mathrm{ord}(a_v) \geqq 0$ for all finite places. We put $\mathfrak{a} = \mathrm{div}(a)$; the positive divisor $\mathfrak{a}$ will sometimes be referred to as the conductor or "the conductor of our problem" ("our problem" being the extension problem described below in Chapter IV).

For every finite place v of k, we define a compact open subgroup $\mathfrak{K}_v$ of G_v by putting

$$\mathfrak{K}_v = \left\{ \begin{pmatrix} u & z \\ a_v w & t \end{pmatrix} \;\middle|\; \begin{matrix} u, z, w, t \in r_v ; \\ ut - a_v wz \in r_v^\times \end{matrix} \right\} ;$$

clearly this depends only upon the conductor $\mathfrak{a}$; it is a maximal compact subgroup of G_v if (and only if) $|a|_v = 1$, i.e. if the place v does not occur in $\mathfrak{a}$.

12. Next, we introduce a quasicharacter α of the idele-class group $k_A^\times/k^\times$, of which we assume that its conductor divides $\mathfrak{a}$. As usual, we write α_v, α_∞ for the quasicharacters induced by α on $k_v^\times$ and on $k_\infty^\times$,

respectively.

Having chosen $\mathfrak{a}$ and α in this manner, we consider functions Φ on G_A, with values in some finite-dimensional space V over $\mathbf{C}$, satisfying the following conditions:

(A) <u>For all</u> $\gamma \in G_k$ <u>and</u> $g \in G_A$, $\Phi(\gamma g) = \Phi(g)$.

(B) <u>For all</u> $g \in G_A$ <u>and</u> $\mathfrak{z} \in \mathfrak{Z}_A$, $\Phi(g\mathfrak{z}) = \Phi(g)\alpha(\mathfrak{z})$.

(C) <u>If</u> v <u>is any finite place outside</u> $\mathfrak{a}$ (i.e. such that $\mathrm{ord}(a_v) = 0$), <u>then, for all</u> $g \in G_A$ <u>and</u> $\mathfrak{p} \in \mathfrak{K}_v$, $\Phi(g\mathfrak{p}) = \Phi(g)$.

(D) <u>If</u> v <u>is a place occurring in</u> $\mathfrak{a}$, <u>and</u>

$$\mathfrak{p} = \begin{pmatrix} u & z \\ a_v w & t \end{pmatrix}$$

<u>is any element of</u> $\mathfrak{K}_v$, <u>then, for all</u> $g \in G_A$, $\Phi(g\mathfrak{p}) = \Phi(g)\alpha_v(t)$.

Remark 1. With the notation of (D), $\mathfrak{p} \longrightarrow \alpha_v(t)$ is a character of $\mathfrak{K}_v$, because of our assumption on the conductor of α; its kernel $\mathfrak{K}'_v$ is an open subgroup of $\mathfrak{K}_v$. If we put $\mathfrak{K}'_v = \mathfrak{K}_v$ for v outside $\mathfrak{a}$, and $\mathfrak{K}' = \prod \mathfrak{K}'_v$, where the product is taken over all the finite places of k, then Φ is constant on cosets $g\mathfrak{K}'$ with respect to $\mathfrak{K}'$. In particular, if k is a function-field, $\mathfrak{K}'$ is an open subgroup of G_A, so that Φ is then locally constant.

Remark 2. If k is a function-field, we always take $V = \mathbf{C}$; if it is a number-field, further conditions will later be imposed upon Φ, and the space V where Φ takes its values will be chosen accordingly (see Chap. IV, §14).

Remark 3. If Φ satisfies conditions (A) to (D), and λ is any quasicharacter of $k_A^\times/k^\times$, the function $g \longrightarrow \Phi(g)\lambda(\det g)$ satisfies similar but somewhat more general conditions; from the point of view adopted here, this would lead to an essentially trivial extension of our theory. From the point of view of representation-theory, however, it corresponds to a non-trivial operation, viz., the tensoring of a given (infinite-dimensional) representation of G_A with the one-dimensional representation $g \longrightarrow \lambda(\det g)$.

13. Let B be the subgroup of $GL(2)$, consisting of all elements of the form $\begin{pmatrix} x & y \\ 0 & 1 \end{pmatrix}$; frequently we shall write (x, y) for the matrix $\begin{pmatrix} x & y \\ 0 & 1 \end{pmatrix}$. The adelized group of B is the subgroup B_A of G_A given by

$$B_A = \left\{ \begin{pmatrix} x & y \\ 0 & 1 \end{pmatrix} \,\middle|\, x \in k_A^\times,\ y \in k_A \right\} .$$

If Φ is the function on G_A introduced in §9, we write F for the function induced by Φ on B_A; in other words, we write

$$F(x, y) = \Phi\left(\begin{pmatrix} x & y \\ 0 & 1 \end{pmatrix}\right) .$$

If Φ is continuous (as will always be assumed) and satisfies at least conditions (A) and (B), it is uniquely determined by F. To see this, we must only show that $G_k B_A \mathfrak{z}_A$ is everywhere dense in G_A. In fact, $B\mathfrak{z}$ is the subgroup of G consisting of the upper triangular matrices; therefore $G/B\mathfrak{z}$ can be identified with the projective line D. Because of general theorems in adele geometry, this implies that $G_A/B_A\mathfrak{z}_A$ can be identified with the adelized line D_A. The image of G_k in D_A is D_k, which is everywhere dense in D_A; this proves our assertion.

Because of assumption (A) on Φ, F is left-invariant with respect to B_k; in other words, it has the following properties:

(a) $F(x, y + \eta) = F(x, y)$ for all $\eta \in k$;

(b) $F(\rho x, \rho y) = F(x, y)$ for all $\rho \in k^\times$.

Because of (a), F has a Fourier expansion:

$$F(x, y) = c_0(x) + \sum_{\xi \in k^\times} c(\xi, x)\psi(\xi y) .$$

In view of (b), we must have, for $\rho \in k^\times$:

$$F(x, y) = c_o(\rho x) + \Sigma c(\xi, \rho x)\psi(\xi\rho y)$$
$$= c_o(\rho x) + \Sigma c(\rho^{-1}\xi, \rho x)\psi(\xi y)$$

and therefore $c_o(x) = c_o(\rho x)$, $c(\xi, x) = c(\rho^{-1}\xi, \rho x)$; in particular, for $\rho = \xi$, we get $c(\xi, x) = c(1, \xi x)$. Now put

$$c(x) = c(1, d^{-1}x)$$

where d is the "differental idele" defined in §7. Then the Fourier series for F appears as

$$F(x, y) = c_o(x) + \sum_{\xi \in k^\times} c(\xi dx)\psi(\xi y) \quad . \tag{4}$$

On the other hand, because of assumptions (C), (D) on Φ, F is right-invariant with respect to the group

$$B_v \cap \mathfrak{K}_v = B_v \cap \mathfrak{K}_v \mathfrak{Z}_v = \{(u, z) \mid u \in r_v^\times, \ z \in r_v\} \tag{5}$$

for every finite place v. In other words, we have:

(a') $F(x, y + z) = F(x, y)$ for all $z \in r_v$;

(b') $F(ux, y) = F(x, y)$ for all $u \in r_v^\times$.

By (a'), we must have, for all $z \in r_v$:

$$c(\xi dx) = c(\xi dx)\psi_v(\xi z) \quad .$$

By the definition of d, this means that $c(\xi dx) = 0$ unless $\operatorname{ord}(\xi dx)_v \geqq 0$. Replacing ξ by 1 and x by $d^{-1}x$, we see that $c(x) = 0$ unless $\mathfrak{m} = \operatorname{div}(x)$ is a positive divisor; taking (b') into account, we see that it depends only upon $\mathfrak{m}$ and x_∞ and may thus be written as $c(x_\infty, \mathfrak{m})$, with $c(x_\infty, \mathfrak{m}) = 0$ unless $\mathfrak{m} \succ 1$. We have thus proved the following:

Proposition 1. *A function* F *on* B_A *is left-invariant under* B_k, *and right-invariant under all groups* $B_v \cap \mathfrak{K}_v$, *if and only if it has a Fourier series of the form* (4), *with* $c_o(\xi x) = c_o(x)$ *for all* $\xi \in k^\times$, $c_o(xu) = c_o(x)$ *for all* $u \in \prod r_v^\times$, c *of the form* $c(x_\infty, \operatorname{div} x)$, *and* $c(x_\infty, \mathfrak{m}) = 0$ *unless* $\mathfrak{m}$ *is positive.*

In terms of F, the coefficients c_o, c are given by the Fourier formulas

$$(6)\quad \begin{aligned} c_o(x) &= \int_{k_A/k} F(x, y)dy \ ; \\ c(x) = c(x_\infty, \operatorname{div} x) &= \int_{k_A/k} F(d^{-1}x, y)\psi(-y)dy \ . \end{aligned}$$

Consequently, if $v \longrightarrow |v|$ is any norm in the space V where Φ and F take their values, we have:

Corollary. *If, for some* $\beta \geqq 0$, *we have* $|F(x, y)| = O(|x|_A^{-\beta})$, *uniformly in* y, *for* $|x|_A \longrightarrow 0$, *then* $|c_o(x)| = O(|x|_A^{-\beta})$, *and there is* $C > 0$ *such that* $|c(x_\infty, \mathfrak{m})| \leqq C|x_\infty|^{-\beta}|\mathfrak{m}|^{-\beta}$ *whenever* $|x_\infty| \leqq 1$.

CHAPTER IV

THE EXTENSION PROBLEM

14. When k is a number-field, further conditions will be imposed on the function Φ of Chapter III, in addition to conditions (A) to (D) of §12. In order to formulate one of these conditions, we define $\mathfrak{K}_w$, for an infinite place w of k, as $\mathfrak{K}_w = O(2, \mathbf{R})$ or $\mathfrak{K}_w = U(2, \mathbf{C})$, according as w is real or imaginary; in both cases it is a maximal compact subgroup of $GL(2, k_w)$. We put $\mathfrak{K}_\infty = \prod \mathfrak{K}_w$, the product being taken over all the infinite places; we also put $\mathfrak{K} = \prod_v \mathfrak{K}_v$, where the product is taken over all the places of k, and the groups $\mathfrak{K}_v$ at the finite places are those defined in Chapter III, §12.

Now we introduce an irreducible representation M_∞ of $\mathfrak{K}_\infty$; as $\mathfrak{K}_\infty$ is compact, its representation-space V is of finite dimension over $\mathbf{C}$. We assume that, <u>on the center of</u> $\mathfrak{K}_\infty$, M_∞ <u>coincides with</u> α, i.e. that $M_\infty(z.1_2) = \alpha_\infty(z).1_V$ whenever $z = (z_w) \in k_\infty^\times$ with $|z_w| = 1$ for all w; α is of course the quasicharacter introduced in §12. If k is a function-field, it should be understood that $\mathfrak{K}_\infty = \{1\}$, M_∞ being then the trivial representation, with $V = \mathbf{C}$.

Now we consider the functions Φ on G_A, with values in V, satisfying conditions (A) to (D) and the following one (which is to be regarded as empty if k is a function-field):

(E) <u>For all</u> $g \in G_A$ <u>and</u> $\mathfrak{k} \in \mathfrak{K}_\infty$, $\Phi(g\mathfrak{k}) = \Phi(g)M_\infty(\mathfrak{k})$.

Because of the assumption on M_∞, this is compatible with (B).

Conditions (B) to (E) prescribe the behavior of Φ under right-translations with respect to the group $\mathfrak{K}\mathfrak{Z}_A$. If we extend M_∞ to a representation M of that group, by putting $M(\mathfrak{z}) = \alpha(\mathfrak{z}).1_V$ for $\mathfrak{z} \in \mathfrak{Z}_A$, $M(\mathfrak{k}) = 1_V$ if v is a finite place not in $\mathfrak{A}$ and $\mathfrak{k} \in \mathfrak{K}_v$, and $M(\mathfrak{k}) = \alpha_v(t).1_V$ if v, $\mathfrak{k}$, t are as in (D), then the conditions (B) to (E) can be replaced by the following:

(B') <u>For all</u> $g \in G_A$, $\mathfrak{k} \in \mathfrak{K}$, $\mathfrak{z} \in \mathfrak{Z}_A$, <u>we have</u> $\Phi(g\mathfrak{k}\mathfrak{z}) = \Phi(g)M(\mathfrak{k}\mathfrak{z})$.

Remark 1. As M_∞ is irreducible, it can be written as the tensor-product $\otimes M_w$ of irreducible representations M_w of the groups $\mathfrak{K}_w$, taken over the infinite places w. The representation-space V is then $\otimes V_w$, where V_w is the representation-space of M_w.

Remark 2. For $k_w = \mathbf{R}$, the "classical case" $k = \mathbf{Q}$ (as discussed in Chapter I) would rather suggest taking for $\mathfrak{K}_w$ the group $SO(2, \mathbf{R})$ of the rotations $r(\theta)$ (cf. §3). Our present choice has been found more convenient for the purposes of the general theory. This creates slight notational discrepancies when one wishes to compare the general theory with the classical one, since the irreducible representations of $O(2, \mathbf{R})$, with only two exceptions, are of degree 2, while those of $SO(2, \mathbf{R})$ are of course of degree 1. No similar question arises for $k_w = \mathbf{C}$, since $U(2, \mathbf{C})\mathfrak{Z}_{\mathbf{C}}$ and $SU(2, \mathbf{C})\mathfrak{Z}_{\mathbf{C}}$ are the same.

15. With that choice of $\mathfrak{K}_w$, we have $G_w = B_w \mathfrak{K}_w \mathfrak{Z}_w$ for all infinite places w; at a finite place v, $B_v \mathfrak{K}_v \mathfrak{Z}_v$ is also the same as G_v whenever v does not occur in $\mathfrak{n}$; if v occurs in $\mathfrak{n}$, $B_v \mathfrak{K}_v \mathfrak{Z}_v$ is at any rate an open subset of G_v (since already $\mathfrak{K}_v$ is such). Thus $B_A \mathfrak{K} \mathfrak{Z}_A$ is an open subset of G_A.

Take any $g \in G_A$; the most elementary approximation theorems show that there is $\gamma \in G_k$ such that $(\gamma^{-1}g)_v$ is in $B_v \mathfrak{K}_v \mathfrak{Z}_v$ for the finitely many places v occurring in $\mathfrak{n}$, and therefore also for <u>all</u> places; this shows that $G_A = G_k B_A \mathfrak{K} \mathfrak{Z}_A$, and therefore that any function Φ, satisfying conditions (A) to (E), is uniquely determined by its values on B_A (without any continuity assumption). A central problem for us, in these lectures, will be to determine when a function F, given on B_A, can be extended to a function Φ on G_A, satisfying conditions (A) to (E) and some further conditions to be stated later.

Proposition 2. A function F on B_A, taking its values in V, can be extended to a function Φ satisfying conditions (A) to (E) if and only if it is left-invariant under B_k, right-invariant under the groups $B_v \cap \mathfrak{K}_v$ for all finite places v, behaves as prescribed by (E) under $B_w \cap \mathfrak{K}_w$ for all infinite places w, and satisfies the following condition:

(I) Put $j = \begin{pmatrix} 0 & 1 \\ -1 & 0 \end{pmatrix}$; then, whenever $b = jb'\mathfrak{p}\mathfrak{z}$ with $b \in B_A$, $b' \in B_A$, $\mathfrak{p} \in \mathfrak{K}$, $\mathfrak{z} \in \mathfrak{Z}_A$, we have $F(b) = F(b')M(\mathfrak{p}\mathfrak{z})$.

Moreover, if this is so, the extension is unique; it is continuous if F is continuous.

Obviously the conditions are necessary. It is easily seen that we have the formula, corresponding to (5) in §13:

$$B_w \cap \mathfrak{K}_w = B_w \cap \mathfrak{K}_w \mathfrak{Z}_w = \{(x, 0) \mid x \in k_w^\times, \ |x|_w = 1\}$$

for all infinite w. Now assume that our conditions for F are satisfied and take any $g \in G_A$; it can be written as $g = \gamma b \mathfrak{p}\mathfrak{z}$, with $\gamma \in G_k$, $b \in B_A$, $\mathfrak{p} \in \mathfrak{K}$, $\mathfrak{z} \in \mathfrak{Z}_A$, and then we must have $\Phi(g) = F(b)M(\mathfrak{p}\mathfrak{z})$. Let it be similarly written as $g = \gamma_1 b_1 \mathfrak{p}_1 \mathfrak{z}_1$. By a well-known elementary theorem (the special case, for $GL(2)$, of "Bruhat's decomposition"), $\gamma^{-1}\gamma_1$ can be written either as $\beta\zeta$ or as $\beta_1 j \beta_2 \zeta$, with β (resp. β_1, β_2) in B_k and $\zeta \in \mathfrak{Z}_k$. In the first case, we have

$$b_2 = b_1^{-1}\beta^{-1}b = \mathfrak{p}_1\mathfrak{z}_1\zeta(\mathfrak{p}\mathfrak{z})^{-1} \in B_A \cap \mathfrak{K}\mathfrak{Z}_A \ ,$$

and our assumptions on F imply that we have

$$\begin{aligned} F(b) = F(\beta b_1 b_2) = F(b_1 b_2) &= F(b_1)M(b_2) \\ &= F(b_1)M(\mathfrak{p}_1\mathfrak{z}_1)M(\mathfrak{p}\mathfrak{z})^{-1} \ . \end{aligned}$$

Similarly, in the second case, we have

$$\beta_1^{-1}b = j(\beta_2 b_1)\mathfrak{p}_1\mathfrak{z}_1\zeta(\mathfrak{p}\mathfrak{z})^{-1} \ ,$$

to which we can apply (I); this gives:

$$F(b) = F(\beta_1^{-1}b) = F(\beta_2 b_1)M(\mathfrak{p}_1\mathfrak{z}_1)M(\mathfrak{p}\mathfrak{z})^{-1} = F(b_1)M(\mathfrak{p}_1\mathfrak{z}_1)M(\mathfrak{p}\mathfrak{z})^{-1} .$$

In both cases we see that the value $F(b)M(\mathfrak{p}\mathfrak{z})$ obtained for $\Phi(g)$ by expressing g in the form $g = \gamma b\mathfrak{p}\mathfrak{z}$ is independent of the choice of that expression: therefore Φ is well defined. Obviously it is uniquely defined, and satisfies conditions (A) and (B'). Finally, if F is continuous, the mapping $b\mathfrak{p}\mathfrak{z} \longrightarrow F(b)M(\mathfrak{p}\mathfrak{z})$ is easily seen to be continuous; therefore the function Φ, defined as above, is continuous on each one of the open subsets $\gamma B_A \mathfrak{K}\mathfrak{Z}_A$ of G_A; as their union is G_A, this shows that Φ is continuous on G_A. This is trivial in the function-field case, since we have seen (§12, Remark 1) that Φ is then, not merely continuous, but locally constant.

16. In view of proposition 2, it is useful to determine when two elements b, b' of B_A satisfy a relation $b = jb'\mathfrak{p}\mathfrak{z}$, with $\mathfrak{p}\mathfrak{z} \in \mathfrak{K}\mathfrak{Z}_A$; clearly this will be so if and only if b_v, b'_v are such that $b_v = jb'_v\mathfrak{p}_v\mathfrak{z}_v$, with $\mathfrak{p}_v\mathfrak{z}_v \in \mathfrak{K}_v\mathfrak{Z}_v$, for all places v of k; in other words, it is a purely local question. Clearly, too, it depends only, for each v, upon the right cosets to which b_v, b'_v belong in B_v with respect to $B_v \cap \mathfrak{K}_v$.

Take first a finite place v; then the latter group is given by (5) in Chapter III, §13; it consists (in the notation explained there) of the matrices (u, z) with $u \in r_v^\times$, $z \in r_v$. Take any element $b = (x, y)$ of B_v; we can write y/x in "reduced form" as e/f with e, f in r_v, $f \neq 0$ and $\sup(|f|_v, |e|_v) = 1$; we can make this unique by prescribing, for instance, that f should be of the form π_v^n with $n \geqq 0$; then $n = \mathrm{ord}(y/x)^-$. Then, putting $t = f^{-1}x$, we have $b = (tf, te)$; we will say that an element b of B_v is written <u>in reduced form</u> if it is so written, and then we call the ideal fr_v of r_v its conductor. Similarly, take any element $b = (x, y)$ of B_A; for each finite place v, write b_v in reduced form as $(t_v f_v, t_v e_v)$; since $x \in k_A^\times$ and $y \in k_A$, we must have $f_v \in r_v^\times$ for almost all places. At an infinite place w, define t_w, f_w, e_w by putting

$$t_w = (x_w \bar{x}_w + y_w \bar{y}_w)^{1/2}, \quad f_w = t_w^{-1} x_w, \quad e_w = t_w^{-1} y_w \ .$$

Then t and f are ideles, e is an adele; the positive divisor $\mathrm{div}(f)$ will be called the conductor of b, and written $\mathrm{cond}(b)$. We will say that (f, e) is reduced if f, e are as above, and that b is written in reduced form if it is written as (tf, te), with (f, e) reduced.

It is now easy to give a complete set of representatives for the right cosets of $B_v \cap \mathfrak{K}_v$ in B_v (resp. of $B_A \cap \mathfrak{K}$ in B_A). In fact, a right translation by an element of $B_v \cap \mathfrak{K}_v$ does not change $|f|_v$ (this is so even when v is infinite) and hence does not change the conductor; if v is finite, it changes e_v only modulo f_v. Therefore, for a finite v, a complete set of representatives is given by the elements $b_v = (t_v \pi_v^n, t_v e_v)$, with $t_v \in k_v^\times$, $n \in \mathbf{N}$, e_v running through a full set of representatives of r_v modulo $\pi_v^n r_v$. For v infinite, we get a complete set by taking $t_w \in k_w^\times$, $f_w \in \mathbf{R}_+^\times$, $e_w \in k_w$.

We can now prove the following result:

Proposition 3. *An element b of B_A can be written as $b = jb'\mathfrak{p}\mathfrak{z}$ if and only if its conductor is a multiple of $\mathfrak{a}$. Then, if b is written in reduced form as (tf, te), we have $b = jb'\mathfrak{p}\mathfrak{z}$ with*

$$b' = (t^{-1} f, t^{-1} e'), \quad \mathfrak{z} = t, \quad \mathfrak{p} = \begin{pmatrix} -e' & -(1+ee')/f \\ f & e \end{pmatrix}$$

where e' is so chosen that $|1 + ee'|_v \leqq |f|_v$ for v finite and $e'_w = -\bar{e}_w$ for w infinite.

One verifies at once that the formulas in question give $b = jb'\mathfrak{p}\mathfrak{z}$, with $\mathfrak{p} \in \mathfrak{K}$ provided e' is chosen as stated; clearly it can be so chosen, since $|e|_v = 1$ unless $|f|_v = 1$. It only remains for us to show that, if $b = jb'\mathfrak{p}\mathfrak{z}$, b and b' have the same conductor, and that this is a multiple of $\mathfrak{a}$. In fact, writing both b and b' in reduced form as (tf, te), $(t'f', t'e')$, we have then

$$b'^{-1}j^{-1}b = t\begin{pmatrix} \frac{-fe'}{f'} & \frac{-1-tt'ee'}{tt'f'} \\ f & e \end{pmatrix} = \mathfrak{p}\mathfrak{z} .$$

In view of the definitions of $\mathfrak{R}$, f, e, this implies that, at all places, $|\mathfrak{z}|_v = |t|_v$ and that the matrix in the middle of the formula must be in $\mathfrak{R}$. But then, at each finite place v, f_v must be a multiple of a_v and also a multiple of f'_v. As $b' = jb(-\mathfrak{p}\mathfrak{z})^{-1}$, one can interchange b and b', so that f'_v must be a multiple of f_v; therefore $|f|_v = |f'|_v \leq |a|_v$.

Corollary. <u>Proposition 2 remains valid if in it we replace (I) by the following:</u>

(I') <u>We have</u> $F(b) = F(b')M(\mathfrak{p}\mathfrak{z})$ <u>whenever the conductor of</u> b <u>is a multiple of</u> $\mathfrak{a}$ <u>and</u> b', $\mathfrak{p}$, $\mathfrak{z}$ <u>are as in proposition</u> 3.

In fact, if at the same time we have $b = jb'_1\mathfrak{p}_1\mathfrak{z}_1$, we must have $b'^{-1}b'_1 \in B_A \cap \mathfrak{R}$; the relation $F(b) = F(b'_1)M(\mathfrak{p}_1\mathfrak{z}_1)$ follows then at once from (I') and the other assumptions made on F in proposition 2.

17. Rather than one function Φ satisfying the conditions (A) to (E), it will be essential for us to consider <u>pairs</u> of functions satisfying similar conditions. These are defined as follows.

The idele $a = (a_v)$ being as in Chapter III, §11, we define an element $\mathfrak{a} = (\mathfrak{a}_v)$ of G_A by putting

$$\mathfrak{a} = \begin{pmatrix} 0 & 1 \\ -a & 0 \end{pmatrix} .$$

If v is finite, and $\mathfrak{p} = \begin{pmatrix} u & z \\ a_v w & t \end{pmatrix}$ is an element of $\mathfrak{R}_v$, we have

$$\mathfrak{a}_v \mathfrak{p} \mathfrak{a}_v^{-1} = \begin{pmatrix} t & w \\ -a_v z & u \end{pmatrix} ;$$

this is also an element of $\mathfrak{k}_v$. The same is true if v is infinite, since then $\mathfrak{a}_v \in \mathfrak{k}_v$. Therefore we have

$$\mathfrak{a}\mathfrak{k}\mathfrak{a}^{-1} = \mathfrak{k} .$$

Now, with each function Φ on G_A, satisfying the conditions (A) to (E), we associate the function Φ' given by

$$\Phi'(g) = \Phi(g\mathfrak{a})\alpha(\det g)^{-1} . \tag{7}$$

It is easily seen that Φ' satisfies conditions (A) to (E) with α replaced by α^{-1} and M replaced by

$$\mathfrak{p} \longrightarrow M'(\mathfrak{p}) = M(\mathfrak{a}^{-1}\mathfrak{p}\mathfrak{a})\alpha_\infty(\det \mathfrak{p})^{-1} .$$

As we have $\det \mathfrak{a} = a$ and $\mathfrak{a}^2 = -a$, we get, by substituting $g\mathfrak{a}$ for g in (7):

$$\Phi(g) = \Phi'(g\mathfrak{a})\alpha(\det g) = \Phi'(g\mathfrak{a})\alpha'(\det g)^{-1} ,$$

so that the relation between Φ and Φ' is symmetrical. The functions Φ, Φ', satisfying the above conditions, will be called an <u>automorphic pair</u>.

As in the case of M, we extend M' to a representation M' of $\mathfrak{k}\mathfrak{z}_A$ in such a way that $\Phi'(g\mathfrak{p}\mathfrak{z}) = \Phi'(g)M'(\mathfrak{p}\mathfrak{z})$ for all $\mathfrak{p} \in \mathfrak{k}$, $\mathfrak{z} \in \mathfrak{z}_A$. This is given by

$$M'(\mathfrak{p}\mathfrak{z}) = M(\mathfrak{a}^{-1}\mathfrak{p}\mathfrak{z}\mathfrak{a})\alpha(\det \mathfrak{p}\mathfrak{z})^{-1} .$$

Let F, F' be the functions induced by Φ, Φ' on B_A. As F determines Φ uniquely, it also determines Φ' and consequently F' uniquely; however, we wish to adopt a different point of view, and, taking F and F' to be given on B_A, to find convenient conditions for them to determine an automorphic pair (Φ, Φ'). A necessary condition for this is the following:

(II) Whenever $b = jb'\mathfrak{p}\mathfrak{z}\mathfrak{a}$ with b and b' in B_A, $\mathfrak{p} \in \mathfrak{K}$, $\mathfrak{z} \in \mathfrak{Z}_A$, we have $F(b) = F'(b')M'(\mathfrak{p}\mathfrak{z})\alpha(\det b'\mathfrak{p}\mathfrak{z})$.

This is obvious, if we write

$$\Phi(b) = \Phi(b'\mathfrak{p}\mathfrak{z}\mathfrak{a}) = \Phi'(b'\mathfrak{p}\mathfrak{z})\alpha(\det b'\mathfrak{p}\mathfrak{z}) \ .$$

In order to apply this, it is useful to do, for the relation $b = jb'\mathfrak{p}\mathfrak{z}\mathfrak{a}$, what proposition 3 did for $b = jb'\mathfrak{p}\mathfrak{z}$. This is as follows:

Proposition 4. An element b of B_A can be written as $b = jb'\mathfrak{p}\mathfrak{z}\mathfrak{a}$ if and only if its conductor is disjoint from $\mathfrak{N}$. Then, if b is written in reduced form as (tf, te), we have $b = jb'\mathfrak{p}\mathfrak{z}\mathfrak{a}$ with

$$b' = (at^{-1}f,\ at^{-1}e'),\quad \mathfrak{z} = -a^{-1}t,\quad \mathfrak{p} = \begin{pmatrix} \frac{1+aee'}{f} & -e' \\ -ae & f \end{pmatrix}$$

where e' is so chosen that $|1+aee'|_v \leqq |f|_v$ for v finite, and $e'_w = -\bar{e}_w$ for w infinite.

We verify the latter part, just as in the proof of proposition 3, by a straightforward calculation. Now assume $b = jb'\mathfrak{p}\mathfrak{z}\mathfrak{a}$; write b, b' in reduced form; we get

$$b'^{-1}j^{-1}b\mathfrak{a}^{-1} = t\begin{pmatrix} \frac{1+tt'ee'}{tt'f'} & \frac{-fe'}{af'} \\ -e & f/a \end{pmatrix} \in \mathfrak{K}\mathfrak{Z}_A \ .$$

Write f/a in "reduced form" as f^*/a^*, with $\sup(|f^*_v|_v, |a^*_v|_v) = 1$ for all finite v (i.e. so that $\mathrm{div}(f^*)$, $\mathrm{div}(a^*)$ are positive and disjoint). We get:

$$b'^{-1}j^{-1}b\mathfrak{a}^{-1} = a^{*-1}t\cdot\begin{pmatrix} a^*\frac{1+tt'ee'}{tt'f'} & \frac{-f^*e'}{f'} \\ -ea^* & f^* \end{pmatrix} .$$

The matrix in the right-hand side must then be in $\mathfrak{K}$; this implies that

$f^*_v e'_v f'^{-1}_v$, hence also $f_v e'_v f'^{-1}_v$, hence $f_v f'^{-1}_v$, must be in r_v for each finite v, so that f is a multiple of f'. But we have $\mathfrak{d}^2 = -a.1_2$, so that our assumption on b, b' can also be written as $b' = jb(\mathfrak{d}\mathfrak{p}^{-1}\mathfrak{d}^{-1})(\mathfrak{z}a)^{-1}\mathfrak{d}$. It is therefore symmetrical in b, b'; this shows that b, b' have the same conductor. But then $f^*_v e'_v f'^{-1}_v$ cannot be in r_v, for a place v occurring in div(f), unless $|f^*|_v = |f|_v$; this implies that v does not then occur in $\mathfrak{a}$.

Just as in the corollary of proposition 3, we conclude at once from this that condition (II) may be replaced by the following one:

(II') *We have* $F(b) = F'(b')M'(\mathfrak{p}\mathfrak{z})\alpha(\det b'\mathfrak{p}\mathfrak{z})$ *whenever the conductor of* b *is disjoint from* $\mathfrak{a}$, *and* b', $\mathfrak{p}$, $\mathfrak{z}$ *are as in proposition* 4.

Here it is understood that F, F' have the behavior prescribed by condition (B') under right-translations by elements of $B_A \cap \mathfrak{k}$.

18. Let S be any set of places of k; we will say that it has the *approximation property* if the projection of $k^\times$ onto the subgroup of $k^\times_A$, consisting of the ideles $x = (x_v)$ such that $x_v = 1$ for all v not in S, is everywhere dense in that group. This is so if and only if there is no non-trivial character of $k^\times_A/k^\times$ which is trivial on $k^\times_v$ for all $v \notin S$. Clearly every finite set of places has the approximation property.

We are now ready to state the main result of this chapter.

Theorem 1. *Let* S *be a set of finite places of* k, *including all the places in* $\mathfrak{a}$; *assume that it has the approximation property (e.g. that it is finite). Let* F, F' *be two functions on* B_A *with values in* V, *left-invariant under* B_k, *and such that*

$$F(bb_1) = F(b)M(b_1), \quad F'(bb_1) = F'(b)M'(b_1)$$

for all $b \in B_A$ *and* $b_1 \in B_A \cap \mathfrak{k}$. *Then* F, F' *can be extended to an automorphic pair* Φ, Φ' *on* G_A *if (and only if) they satisfy condition* (II) *of* §17 *whenever the common conductor of* b *and* b' *in* (II) *is disjoint from* S.

We first observe that our assumption is symmetrical (as it should be) in F and F'; in fact, as observed before, the relation $b = jb'\mathfrak{p}\mathfrak{z}\mathfrak{a}$ can be rewritten as $b' = jb\mathfrak{p}'\mathfrak{z}'\mathfrak{a}$, with $\mathfrak{p}' = \mathfrak{a}\mathfrak{p}^{-1}\mathfrak{a}^{-1}$, $\mathfrak{z}' = \mathfrak{z}^{-1}a^{-1}$, and it is easily verified that we have

$$M(\mathfrak{p}'\mathfrak{z}')\alpha(\det b\mathfrak{p}'\mathfrak{z}')^{-1} = M'(\mathfrak{p}\mathfrak{z})^{-1}\alpha(\det b'\mathfrak{p}\mathfrak{z})^{-1} .$$

Our next step will now be to show that, under our assumptions, F satisfies condition (I') of the corollary of proposition 3, §16; then F' satisfies it too, by symmetry.

Take b and b' as in that corollary, i.e. as in proposition 3 of §16; we first prove a lemma:

Lemma 1. <u>With</u> b, b' <u>as above, there is</u> $\lambda \in k^{\times}$ <u>such that the conductors of the elements</u> $(1, \lambda)b$ <u>and</u> $(-\lambda, 1)b'$ <u>of</u> B_A <u>are both disjoint from</u> S.

In fact, we have

$$(1, \lambda)b = (tf, \lambda + te) ,$$

and to say that its conductor is disjoint from S is to say that, for all $v \in S$, $|(\lambda + te)/tf|_v \leqq 1$. Similarly

$$(-\lambda, 1)b' = (-\lambda t^{-1}f, 1 - \lambda t^{-1}e') ,$$

whose conductor is disjoint from S if and only if $|\lambda^{-1}tf^{-1} - e'f^{-1}|_v \leqq 1$ for $v \in S$. Now, if $|f|_v = 1$, both conditions are satisfied provided $|\lambda/t|_v = 1$. On the other hand, let v be one of the finitely many places in S for which $|f|_v < 1$; then, if $\lambda = -t_v e_v(1 + f_v z)$ with $z \in r_v$, we have

$$\lambda^{-1}t_v f_v^{-1} - e'_v f_v^{-1} = \frac{1 - e_v e'_v(1 + f_v z)}{e_v f_v(1 + f_v z)} ,$$

and this is in r_v because of the choice of e' in proposition 3. Now the approximation property for S says that there is $\lambda \in k^\times$ satisfying these conditions. This proves lemma 1.

It is now easy to verify (I'); b and b' being as above, put $\beta = (1, \lambda)$, $\beta' = (-\lambda, 1)$. Since S contains all the places in $\mathfrak{a}$, we can apply proposition 4 to βb and to $\beta' b'$, so that we can write $\beta b = j b_1 \mathfrak{p}_1 \mathfrak{z}_1 \mathfrak{a}$, $\beta' b' = j b_2 \mathfrak{p}_2 \mathfrak{a}$; combining these with $b = j b' \mathfrak{p}\mathfrak{z}$, we get

$$b_2 = (j^{-1}\beta' j^{-1} \beta^{-1} j) b_1 \mathfrak{p}_1 \mathfrak{z}_1 \mathfrak{a} (\mathfrak{p}\mathfrak{z})^{-1} (\mathfrak{p}_2 \mathfrak{z}_2 \mathfrak{a})^{-1} = \beta'' b_1 \mathfrak{p}' \mathfrak{z}'$$

where we have put

$$\beta'' = (-\lambda, -1), \ \mathfrak{p}'\mathfrak{z}' = \mathfrak{p}_1 \mathfrak{z}_1 . \mathfrak{a} \mathfrak{p}^{-1} \mathfrak{z}^{-1} \mathfrak{a}^{-1} . (\mathfrak{p}_2 \mathfrak{z}_2)^{-1} \ .$$

Now, applying (II), we have:

$$F(b) = F(\beta b) = F'(b_1) M'(\mathfrak{p}_1 \mathfrak{z}_1) \alpha(\det b_1 \mathfrak{p}_1 \mathfrak{z}_1) \ ,$$

$$F(b') = F(\beta' b') = F'(b_2) M'(\mathfrak{p}_2 \mathfrak{z}_2) \alpha(\det b_2 \mathfrak{p}_2 \mathfrak{z}_2) \ .$$

At the same time, in the relation between b_1 and b_2, we have $\beta'' \in B_k$, so that $\mathfrak{p}'\mathfrak{z}' \in B_A \cap \mathfrak{k}\mathfrak{z}_A$; in view of our assumptions on F', we have therefore:

$$\begin{aligned} F'(b_1) = F'(\beta'' b_1) &= F'(b_2) M'(\mathfrak{p}'\mathfrak{z}')^{-1} \\ &= F'(b_2) M'(\mathfrak{p}_2 \mathfrak{z}_2) M'(\mathfrak{a}\mathfrak{p}\mathfrak{z}\mathfrak{a}^{-1}) M'(\mathfrak{p}_1 \mathfrak{z}_1)^{-1} \ . \end{aligned}$$

Combining these, and taking into account the relationship between M and M', we get at once $F(b) = F(b') M(\mathfrak{p}\mathfrak{z}) \alpha(z)$, where we have put

$$\begin{aligned} z &= \det(b_2 \mathfrak{p}_2 \mathfrak{z}_2)^{-1} \det(\mathfrak{p}\mathfrak{z})^{-1} \det(b_1 \mathfrak{p}_1 \mathfrak{z}_1) \\ &= \det(j^{-1} \beta' b' \mathfrak{a}^{-1})^{-1} \det(b^{-1} j b') \det(j^{-1} \beta b \mathfrak{a}^{-1}) \ . \end{aligned}$$

This gives $z = -\lambda^{-1}$, hence $\alpha(z) = 1$ since α is trivial on $k^{\times}$, which completes the verification of (I') for F.

Now the corollary of proposition 3, §16, shows that F can be extended to a function Φ on G_A, satisfying conditions (A), (B'), and similarly that F' can be extended to a function Φ' satisfying the same conditions with α^{-1}, M' replacing α, M. Put now $\Phi'_1(g) = \Phi(g\mathfrak{a})\alpha(\det g)^{-1}$, and call F'_1 the function induced by Φ'_1 on B_A. In view of what had been proved earlier, the pair F, F'_1 must also satisfy condition (II). In view of proposition 4, this implies that $F'_1(b') = F'(b')$ whenever the conductor of b' is disjoint from S. Now take any b in B_A; the same construction carried out in the proof of lemma 1 (but applied to b alone) shows that we can choose $\lambda \in k$ so that $b' = (1, \lambda)b$ has a conductor disjoint from S. From $F'_1(b') = F'(b')$, we conclude now $F'_1(b) = F'(b)$. Therefore $F' = F'_1$, $\Phi' = \Phi'_1$, and Φ, Φ' are an automorphic pair.

I have been unable to answer the following question: $\mathfrak{a}$ being given, can one find a <u>finite</u> set of places S' such that theorem 1 remains valid when one takes for S the complement of S' ? If true, this would require a different proof, since S cannot then have the approximation property.

CHAPTER V

THE CONVERGENCE LEMMAS

19. We will need some lemmas on the convergence of the Fourier series (4) of Chapter III, §13, and the order of magnitude of F in terms of that of c. It will be convenient to discuss this separately for function-fields and for number-fields.

Take first the case $\mathrm{char}(k) > 1$, i.e. let k be a function-field. Notations being as in Chapter III, proposition 1 shows that c can be written as $c(\mathfrak{m})$, with $\mathfrak{m} = \mathrm{div}(x)$; as this is 0 unless $\mathfrak{m}$ is positive, we can then introduce the formal Dirichlet series $\Sigma c(\mathfrak{m})|\mathfrak{m}|^s$; moreover, if F satisfies the assumption in the corollary of proposition 1, this is absolutely convergent for $\mathrm{Re}(s) > \beta+1$, so that it is "a Dirichlet series belonging to k" (as defined in §6); it will be called <u>the Dirichlet series attached to</u> F, or to Φ.

We will now show that, in the case $\mathrm{char}(k) > 1$, the convergence of the Fourier series for F is trivial. More generally, we have:

Lemma 2. *For* $\mathrm{char}(k) > 1$, *let* $\mathfrak{m} \longrightarrow c(\mathfrak{m})$ *be any mapping of* $\mathfrak{M}_+$ *into* $\mathbb{C}$; *put* $c(\mathfrak{m}) = 0$ *when* $\mathfrak{m}$ *is not positive. Then the series*

$$\sum_{\xi \in k^\times} c(\mathrm{div}\ \xi x)\psi(\xi y)$$

is trivially convergent, and uniformly so over every compact subset of B_A; *it is identically* 0 *for* $|x| > 1$. *If* $|c(\mathfrak{m})| \leqq C|\mathfrak{m}|^{-\alpha}$ *with* $C > 0$ *and* $\alpha \geqq 0$, *there is* $C' > 0$ *such that, for all* x:

$$\sum_{\xi \in k^\times} |c(\mathrm{div}(\xi x))| \leqq C'|x|^{-\alpha-1} .$$

Put $\mathfrak{m} = \mathrm{div}(x)$ and $m = \deg(\mathfrak{m})$, so that $|x| = |\mathfrak{m}| = q^{-m}$, where q is the number of elements of the constant field for k. In the given series, only those terms are $\neq 0$ which correspond to the elements ξ of $k^{\times}$ for which $\mathrm{div}(\xi) \succ \mathfrak{m}^{-1}$; for each x, they are in finite number (which is what was meant by saying that the series is "trivially convergent"). There is no such term if $m < 0$, i.e. $|x| > 1$; otherwise, by the theorem of Riemann-Roch, the number of such terms is $< q^{m+1}$; if these terms are all $\leqq C q^{-ma}$, as assumed in the last assertion of the lemma, we get the conclusion of that assertion, with $C' = Cq$. Let K be a compact subset of $k_A^{\times}$; as the mapping $x \longrightarrow \mathrm{div}(x)$ of $k_A^{\times}$ onto $\mathfrak{M}$ is locally constant, there is a divisor $\mathfrak{m}_0$ such that $\mathrm{div}(x) \prec \mathfrak{m}_0$ for all $x \in K$; therefore, for $x \in K$, only those terms of our series for which $\mathrm{div}(\xi) \succ \mathfrak{m}_0^{-1}$ can be $\neq 0$; therefore the series is uniformly convergent (and even, in an obvious sense, "uniformly trivially convergent") for $x \in K$.

20. Before we deal with number-fields, we need a preliminary lemma.

Lemma 3. *Let* E *be a vector-space of dimension* n *over* $\mathbf{R}$; *let* N *be a norm in* E, *and* L *a lattice in* E. *Then, for every* $\lambda > 0$, *there are constants* $\mu > 0$, $C > 0$, $C' > 0$ *such that*

$$\sum_{\substack{e \in L \\ e \neq 0}} \exp(-\lambda t N(e)) \leqq C t^{-n} \quad \textit{for} \quad 0 < t \leqq 1 \,,$$
$$\leqq C' \exp(-\mu t) \quad \textit{for} \quad t \geqq 1 \,.$$

Let $(e_1, \ldots, e_n)$ be a basis for L; then the formula

$$N'\left(\sum_{i=1}^{n} x_i e_i\right) = \sum_{i=1}^{n} |x_i| \,,$$

for $(x_1, \ldots, x_n) \in \mathbf{R}^n$, defines a norm in E, and there is $\rho > 0$ such that $N' \leqq \rho N$. Put $\mu = \lambda/\rho$; calling S the sum in the left-hand side of our inequality, we have

$$S \leq \sum_{\substack{e \in L \\ e \neq 0}} \exp(-\mu t N'(e)) = \sum_{\substack{m \in Z^n \\ m \neq 0}} \exp(-\mu t \Sigma |m_i|)$$

$$= \left(\sum_{\nu=-\infty}^{+\infty} e^{-\mu t \nu} \right)^n - 1 = \left(\frac{1 + e^{-\mu t}}{1 - e^{-\mu t}} \right)^n - 1 .$$

Our conclusion follows from this at once.

21. Lemma 4. *Let k be a number-field, of degree d over $\mathbf{Q}$. For each infinite place w of k, let φ_w be a continuous function, > 0 on $\mathbf{R}_+^\times$, such that $\varphi_w(p) = O(p^{-A})$, with some $A \geq 0$, for $p \longrightarrow 0$, and $\varphi_w(p) = O(e^{-\lambda p})$, with some $\lambda > 0$, for $p \longrightarrow +\infty$. Let c be a mapping of $k_\infty^\times \times \mathfrak{M}$ into $\mathbf{C}$, such that $c(x_\infty, \mathfrak{m}) = 0$ unless $\mathfrak{m}$ is positive, and that*

$$|c(x_\infty, \mathfrak{m})| \leq C \prod_w \varphi_w(\mathrm{abs}\ x_w) . |\mathfrak{m}|^{-\alpha} ,$$

with $C > 0$, $\alpha \geq 0$, for all x_∞, $\mathfrak{m}$. Then there are constants $\beta \geq 0$, $\mu > 0$, $C' > 0$, $C'' > 0$ such that, for all $x \in k_A^\times$:

$$\sum_{\xi \in k^\times} |c(\xi x_\infty, \mathrm{div}(\xi x))| \leq C' |x|^{-\beta} \text{ if } |x| \leq 1 ,$$
$$\leq C'' \exp(-\mu |x|^{1/d}) \text{ if } |x| \geq 1 ;$$

and the Fourier series

$$\sum_{\xi \in k^\times} c(\xi x_\infty, \mathrm{div}(\xi x)) \psi(\xi y)$$

is uniformly absolutely convergent over compact subsets of B_A.

In the assumption on c, it should be understood that abs x means the "ordinary" absolute value, i.e. $|x|$ if $k_w = \mathbf{R}$, and $(x\bar{x})^{1/2}$ if $k_w \in \mathbf{C}$. If that assumption is satisfied for some $\alpha \geq 0$, it remains so if we substitute for α any $\alpha' > \alpha$; therefore we may assume that $\alpha \geq A$. We will denote various constants by C_1, C_2, etc.

If we write k_A^1 for the kernel of $x \longrightarrow |x|$ in $k_A^\times$, it is well-known that $k_A^1/k^\times$ is compact; therefore we can choose a compact subset K of k_A^1 so that, whenever $|x| = 1$, there is $\xi \in k^\times$ such that $\xi^{-1}x \in K$. Now take any $x \in k_A^\times$; put $p = |x|^{1/d}$; let z be the idele given by $z_w = p$ for w infinite, $z_v = 1$ for v finite; then $|z| = p^d$, so that $|z^{-1}x| = 1$; therefore we may write $z^{-1}x = \xi_1 x'$ with $x' \in K$, $\xi_1 \in k$. In the inequality to be proved, nothing is changed if we replace x by $\xi_1^{-1}x$; therefore it will be enough if we prove that inequality for $x = zx'$ with z as above and $x' \in K$. In order to prove also the last assertion in our conclusion, we assume merely $x' \in K'$, where K' is any compact subset of k_A^1.

For x, z, x' as above, we have $\mathrm{div}(x) = \mathrm{div}(x')$. As in §19, we observe that $x \longrightarrow \mathrm{div}(x)$ is locally constant; therefore, for $x' \in K'$, there is a positive divisor $\mathfrak{m}$ such that $\mathrm{div}(x) \prec \mathfrak{m}$ for all z; then, in the series under consideration, only those terms for which $\mathrm{div}(\xi) \succ \mathfrak{m}^{-1}$ can be $\neq 0$; these are the terms corresponding to the elements $\xi \neq 0$ of the fractional ideal $\mathfrak{m}^{-1}$. Put $|\xi|_\infty = \prod |\xi|_w$, the product being taken over the infinite places; we have

$$1 = |\xi| = |\xi|_\infty \cdot |\mathrm{div}(\xi)| \;,$$

and therefore, for $x' \in K'$:

$$|\mathrm{div}(\xi x)| = |\xi|_\infty^{-1} \cdot |\mathrm{div}(x')| \geqq C_1 \cdot |\xi|_\infty^{-1} \;.$$

Applying our assumption on c, we get:

$$|c(\xi x_\infty, \mathrm{div}(\xi x))| \leqq C_2 \cdot |\xi|_\infty^{a} \prod_w \varphi_w(p\, \mathrm{abs}(x'_w \xi_w)) \;,$$

where ξ_w is, of course, the image of ξ in k_w under the natural embedding of k into k_w.

For $p > 0$, define φ'_w by $\varphi'_w(p) = p^a \varphi_w(p)$ for w real and $\varphi'_w(p) = p^{2a} \varphi_w(p)$ for w imaginary; put $\varphi'_w(0) = 0$ for all w.

Expressing φ_w in terms of φ'_w in the above inequality, we get (since (x'_w) lies in a compact subset of $k_w^\times$):

$$|c(\xi x_\infty, \operatorname{div}(\xi x))| \leqq C_3 . p^{-d\alpha} \prod_w \varphi'_w(p \operatorname{abs}(x'_w \xi_w)) \ .$$

Take any λ' such that $0 < \lambda' < \lambda$; in view of our assumptions on φ_w, and of our assumption $\alpha \geqq A$, we have, for all $p \geqq 0$, $\varphi'_w(p) \leqq C_4 \exp(-\lambda' p)$. Now, considering $k_\infty = \prod k_w$ as a vector-space of dimension d over **R**, define a norm in that space by putting, for $z = (z_w) \in k_\infty$:

$$N(z) = \inf_{x' \in K'} \sum_w \operatorname{abs}(x'_w z_w) \ .$$

We have now shown that the sum, in the inequality to be proved, is

$$\leqq C_5 \, p^{-d\alpha} \sum_{\substack{\xi \in \mathfrak{m}^{-1} \\ \xi \neq 0}} \exp(-\lambda' p N(\xi)) \ .$$

Applying lemma 3 to the sum in the right-hand side, we get the result announced in our lemma, as well as the uniform convergence of the given sum for x (and not merely x') lying in a compact subset of $k_A^\times$.

CHAPTER VI

HECKE OPERATORS

22. As we have observed in §8, $\mathfrak{K}_v$ is a maximal compact subgroup of G_v, at a finite place v of k, if (and only if) v does not occur in the "conductor" $\mathfrak{n}$. Take any such place v; as explained in §13, we write $(\pi_v, 0)$ for the matrix

$$\begin{pmatrix} \pi_v & 0 \\ 0 & 1 \end{pmatrix} .$$

In G_v, we consider the double coset

$$H_v = \mathfrak{K}_v . (\pi_v, 0). \mathfrak{K}_v \ ;$$

from the theory of "elementary divisors" in G_v, it follows that H_v is no other than the set of matrices in G_v with coefficients in r_v and with a determinant $\delta \in \pi_v . r_v^\times$; it does not depend upon the choice of the prime element π_v in k_v. From this, one concludes easily that H_v can be written as follows as a disjoint union of right $\mathfrak{K}_v$-cosets:

$$H_v = \bigcup_u (\pi_v, u). \mathfrak{K}_v \cup \begin{pmatrix} 1 & 0 \\ 0 & \pi_v \end{pmatrix} . \mathfrak{K}_v \ ,$$

where u runs through a complete set of representatives of r_v modulo π_v; as usual, we write $q_v = |\pi_v|^{-1}$ for the number of such representatives (the "module" of k_v).

Now take any function Φ on G_A which is right-invariant under $\mathfrak{K}_v$, and consider the function

$$\Psi(g) = \int_{H_v} \Phi(gh)dh \ ,$$

where dh is the Haar measure on G_v, normalized by taking $\mathrm{mes}(\mathfrak{K}_v) = 1$. Clearly this is also right-invariant under $\mathfrak{K}_v$; moreover, in view of the right coset decomposition given above for H_v, it can be written as:

$$\Psi(g) = \sum_u \Phi(g.(\pi_v, u)) + \Phi(g.\begin{pmatrix} 1 & 0 \\ 0 & \pi_v \end{pmatrix}) \quad . \tag{8}$$

It is also obvious that, if Φ satisfies all the conditions (A) to (E), so does Ψ. _We will write_ T_v _for the operator_ $\Phi \longrightarrow \Psi$, _restricted to the functions satisfying_ (A) _to_ (E) _on_ G_A, and call this _the Hecke operator at_ v. Thus we have attached a Hecke operator to each finite place v of k, _not occurring in the conductor_ $\mathfrak{n}$. It is also possible to define Hecke operators for the places occurring in $\mathfrak{n}$, but the above definition would have to be somewhat modified; as this is not necessary for our purposes in these lectures, we omit it.

23. It is clear that the Hecke operators T_v, T_w at any two places v, w of k (not occurring in $\mathfrak{n}$) commute with each other. We denote by $T_v\Phi$ the image of Φ under T_v; moreover, if F is the function induced by Φ on B_A, we will write T_vF for the function induced by $T_v\Phi$ on B_A; this is justified, not merely by the fact that Φ is uniquely determined by F, but, more precisely, by the following:

Proposition 5. _Let_ F _be given by its Fourier series (as in_ (4), §13). _Then_ T_vF _has the Fourier series of the same form, with coefficients_ c'_o, c' _given by_

$$c'_o(x) = q_v c_o(x\pi_v) + \alpha_v(\pi_v) c_o(x\pi_v^{-1}) \quad ,$$

$$c'(x) = q_v c(x\pi_v) + \alpha_v(\pi_v) c(x\pi_v^{-1}) \quad \underline{\text{if}} \quad \mathrm{ord}(x_v) \geqq 0$$

$$c'(x) = 0 \quad \underline{\text{if}} \quad \mathrm{ord}(x_v) < 0 \quad .$$

In fact, in (8), replace g by (x, y); using condition (B) for Φ, we get:

$$(9) \qquad (T_v F)(x, y) = \sum_u F(x\pi_v, y + xu) + \alpha_v(\pi_v) F(x\pi_v^{-1}, y) .$$

Here replace F by its Fourier series; u is to be understood as the adele with the component u at place v, and 0 at all other places. Taking into account the fact that, for every x such that $\mathrm{ord}(d_v x_v \pi_v) \geqq 0$, the sum

$$\sum_u \psi_v(x_v u)$$

has the value q_v or 0 according as $\mathrm{ord}(d_v x_v) \geqq 0$ or not, we get the result stated above.

Here one should recall that the conductor of α was assumed (in §12) to be a divisor of $\mathfrak{m}$; in particular, as v does not occur in $\mathfrak{m}$, it does not occur in the conductor of α, so that $\alpha_v(\pi_v)$ is (as it should) independent of the choice of the prime element π_v in k_v. Instead of $\alpha_v(\pi_v)$, we may also write $\alpha(\mathfrak{y})$ if $\mathfrak{y}$ is the divisor corresponding to the place v.

24. As was first discovered by Hecke, the eigenfunctions of the Hecke operators T_v play a specially important role in the theory of automorphic functions on G_v; proposition 5 makes it possible to interpret this condition in terms of the coefficients of the Fourier series for F. In order to do this, define a sequence $(\gamma_n)_{n \geqq 0}$, for a given value of $\alpha(\mathfrak{y})$ and a given $\lambda \in \mathbf{C}$, by the formal power-series expansion:

$$(10) \qquad \left(1 - \frac{\lambda}{q_v} T + \frac{\alpha(\mathfrak{y})}{q_v} T^2\right)^{-1} = \sum_{n=0}^{\infty} \gamma_n T^n \quad ;$$

also, put $\gamma_{-1} = 0$ and $\gamma_{-n-1} = - q_v^n \alpha(\mathfrak{y})^{-n} \gamma_{n-1}$ for $n \geqq 1$. With this notation, we have:

Proposition 6. F <u>is an eigenfunction of</u> T_v <u>for the eigenvalue</u> λ <u>if and only if the coefficients</u> c_o, c <u>of its Fourier series satisfy the following conditions:</u>

(a) <u>for</u> $\mathrm{ord}(x_v) = 0$ <u>and all</u> $n \geq 0$, $c(x\pi_v^n) = \gamma_n c(x)$;

(b) <u>for all</u> n <u>and all</u> x:

$$c_o(x\pi_v^n) = \gamma_n c_o(x) + \gamma_{n+1} c_o(x\pi_v^{-1}) ,$$

<u>where the</u> γ_n <u>are defined as above in terms of</u> q_v, $\alpha(\mathfrak{y})$, λ.

In fact, in view of proposition 5, the condition $T_v F = \lambda F$ is equivalent to a difference equation for $n \longrightarrow c(x\pi_v^n)$ and another one for $n \longrightarrow c_o(x\pi_v^n)$; this can be solved at once in terms of the γ_n, with the result stated in proposition 6.

Consider in particular the case when k is a function-field; then we can write $c(x) = c(\mathfrak{m})$ with $\mathfrak{m} = \mathrm{div}\, x$, and condition (a) in proposition 3 is nothing else than the Euler property for the Dirichlet series $\Sigma\, c(\mathfrak{m})\,|\mathfrak{m}|^s$, with the Euler factor

$$(1 - \lambda q_v^{-1-s} + \alpha(\mathfrak{y}) q_v^{-1-2s})^{-1}$$

since $|\mathfrak{y}| = q_v^{-1}$.

25. As the operators T_v, attached to the finite places v outside $\mathfrak{A}$, commute with one another, we can obviously extend their definition to a semigroup of operators $T_{\mathfrak{n}}$, corresponding to the positive divisors $\mathfrak{n}$ disjoint from $\mathfrak{A}$, by prescribing that $\mathfrak{n} \longrightarrow T_{\mathfrak{n}}$ shall be a morphism; in other words, if $\mathfrak{n} = \prod \mathfrak{y}_v^{n(v)}$, we put $T_{\mathfrak{n}} = \prod (T_v)^{n(v)}$. This, of course, can further be extended to an algebra consisting of all finite sums $\Sigma\, a(\mathfrak{n}) T_{\mathfrak{n}}$.

Now let (Φ, Φ') be an automorphic pair in the sense of Chapter III, §17, and let v be any finite place outside $\mathfrak{A}$. Then $\mathfrak{a}_v$ is in $\mathfrak{R}_v$, so that, if H_v is the double coset defined in §22, we have

$H_v \mathfrak{a}_v = \mathfrak{a}_v H_v = H_v$ and $H_v \mathfrak{a} = \mathfrak{a} H_v$. Applying now the definition of T_v to Φ', as given by (7) of §17, we get

$$\begin{aligned} T_v \Phi'(g) &= \int_{H_v} \Phi'(gh)dh = \int_H \Phi(gh\mathfrak{a})\alpha(\det gh)^{-1}dh \\ &= \alpha(\pi_v \det g)^{-1} \int_{H_v} \Phi(g\mathfrak{a}h)dh \\ &= \alpha(\mathfrak{y})^{-1}(T_v\Phi)'(g) \ . \end{aligned}$$

In other words, $T_v\Phi$ and $\alpha(\mathfrak{y})T_v(\Phi')$ make up an automorphic pair. From this it follows at once that, if $\mathfrak{n}$ is any positive divisor disjoint from $\mathfrak{A}$, $T_{\mathfrak{n}}\Phi$ and $\alpha(\mathfrak{n})T_{\mathfrak{n}}(\Phi')$ make up an automorphic pair.

In particular, if Φ is an eigenfunction of $T_{\mathfrak{n}}$ for the eigenvalue λ, Φ' is one for the eigenvalue $\lambda\alpha(\mathfrak{n})^{-1}$. If $\mathfrak{n}$ is as before the prime divisor $\mathfrak{y}$, this gives the following Euler factor, in the function-field case, for the Dirichlet series attached to Φ':

$$(1 - \lambda\alpha(\mathfrak{y})^{-1} q_v^{-1-s} + \alpha(\mathfrak{y})^{-1} q_v^{-1-2s})^{-1} \ .$$

CHAPTER VII

FUNCTION-FIELDS

26. All the tools are now at hand for dealing fully with the function-field case, which in several respects is simpler than the case of characteristic 0. This will be done now, at the cost of some repetition later on.

We start with two $\mathbf{C}$-valued functions F, F' on B_A, left-invariant under B_k and right-invariant under $B_A \cap \mathfrak{K}$; according to proposition 1 of Chapter III, §13, we may assume that they are given by their Fourier series and write the coefficients of these series as $c_o(\mathfrak{m})$, $c(\mathfrak{m})$ and $c'_o(\mathfrak{m})$, $c'(\mathfrak{m})$, respectively, with $c(\mathfrak{m}) = 0$ and $c'(\mathfrak{m}) = 0$ unless $\mathfrak{m}$ is positive, and $c_o(\xi\mathfrak{m}) = c_o(\mathfrak{m})$, $c'_o(\xi\mathfrak{m}) = c'_o(\mathfrak{m})$ for all $\xi \in k^\times$.

Now assume of F, F' that they satisfy, for some $\beta \geqq 0$, the condition in the corollary of proposition 1, §13, i.e. $F(x, y) = O(|x|_A^{-\beta})$ for $|x|_A \longrightarrow 0$, uniformly in y; then that corollary shows that the extended Dirichlet series

$$Z(\omega) = \Sigma c(\mathfrak{m})\omega(\mathfrak{m}), \quad Z'(\omega) = \Sigma c'(\mathfrak{m})\omega(\mathfrak{m})$$

are absolutely convergent somewhere. For each integer $n \geqq 0$, we also introduce the partial sums $Z_n(\omega)$, $Z'_n(\omega)$, consisting of the finitely many terms for which $\deg(\mathfrak{m}) = n$ in the series for $Z(\omega)$ and for $Z'(\omega)$, respectively. For $n < 0$, we put $Z_n(\omega) = 0$, $Z'_n(\omega) = 0$.

We can apply theorem 1 of Chapter IV, §18, to F and F'; accordingly, if F and F' are the functions induced on B_A by an automorphic pair (Φ, Φ'), they satisfy condition (II), and a fortiori condition (II') of Chapter IV, §17. As observed there, we also know that (II') implies (II), so that conversely, if F and F' satisfy (II') whenever the conductor of b in (II') is disjoint from the set S in theorem 1, F and F' can be extended to an automorphic pair. We also observe that, if b and b' are as in (II'), i.e. if they are as in proposition 4 of Chapter IV, §17, we have, in view of the definition of b', $\mathfrak{p}$, $\mathfrak{z}$ in that proposition, and of that of M' in §17:

$$\det \mathfrak{p} = 1, \quad \det(b'\mathfrak{z}) = -f\mathfrak{z} \; ,$$

$$M'(\mathfrak{p}\mathfrak{z}) = M'(\mathfrak{p})\alpha(\mathfrak{z})^{-1} = \prod_{v/\mathfrak{a}} \alpha_v(f_v)^{-1} \cdot \alpha(\mathfrak{z})^{-1} \; ,$$

where the product is taken over the places v in $\mathfrak{a}$. At the same time, since $\mathrm{div}(f)$ is disjoint from $\mathfrak{a}$, we have, in view of the definitions in Chapter II, §8:

$$\alpha(f) = \alpha(\mathrm{div}\, f) \prod_{v/\mathfrak{a}} \alpha_v(f_v) \; .$$

Put $\mathfrak{f} = \mathrm{div}\, f$. The conclusion of (II') may now be rewritten as follows:

$$F(tf, te) = F'(at^{-1}f, at^{-1}e')\alpha(\mathfrak{f}) \; ,$$

where $\mathfrak{f}$ is disjoint from $\mathfrak{a}$ (and even from the set S, if one makes use of theorem 1), and e, e' are such that $|1 + aee'|_v \leq |f|_v$ for all v. Replacing F and F' by their Fourier series, we get

$$\text{(11)} \quad \begin{aligned} & c_o(tf) + \sum_{\xi \in k^\times} c(\xi d t f)\psi(\xi t e) \\ & -\alpha(\mathfrak{f})[c'_o(at^{-1}f) + \sum_{\xi \in k^\times} c'(\xi d a t^{-1} f)\psi(\xi a t^{-1} e')] = 0 \; . \end{aligned}$$

As observed earlier, no question of convergence arises, since the series are really finite sums. Replace t by tu, where $|u| = 1$; as a function of u, the left-hand side does not change if one replaces u by ξu with $\xi \in k^\times$, and may therefore be regarded as a function on the compact group $k_A^1/k^\times$, where k_A^1 is the kernel of $x \longrightarrow |x|$ in $k_A^\times$. As such, it is 0 if and only if all its Fourier coefficients are 0; this means that, multiplying with an arbitrary character ω of $k_A^1/k^\times$ (or, what amounts to the same, with an arbitrary quasicharacter of $k_A^\times/k^\times$) and integrating over that compact group, one gets 0. Write now:

$$(12)\qquad \begin{aligned} I(f, e, t, \omega) &= \int_{k_A^1/k^\times} [F(tuf, tue) - c_o(tuf)]\omega(u)d^\times u \\ &= \int_{k_A^1/k^\times} [\Sigma_{\xi\in k^\times} c(\xi dtuf)\psi(\xi tue)]\omega(u)d^\times u \end{aligned}$$

where $d^\times u$ is a Haar measure (which will be suitably normalized later on). Call $I'(f, e, t, \omega)$ the similar integral where F, c, c_o are replaced by F', c', c'_o; then we see that (11) is equivalent to

$$(13)\qquad \begin{aligned} &I(f, e, t, \omega) - \alpha(\mathfrak{f})I'(f, e', at^{-1}, \omega^{-1}) \\ &\quad + \int_{k_A^1/k^\times} [c_o(tuf) - \alpha(\mathfrak{f})c'_o(at^{-1}u^{-1}f)]\omega(u)d^\times u = 0 \ . \end{aligned}$$

In (12), it is understood that $I(f, e, t, \omega)$ is defined only if (f, e) is reduced, i.e. if we have $\sup(|f|_v, |e|_v) = 1$ for all places v, and also if $\mathfrak{f} = \mathrm{div}(f)$ is disjoint from $\mathfrak{A}$. From the properties of F, and of c_o, described in Chapter III, it follows at once that $I(f, e, t, \omega)$ is not changed if we replace f by any idele f_1 with the same divisor, or if we replace e by $e + fs$ with $s_v \in r_v$ for all v; this can also be seen at once from the second integral in (12), and the properties of c.

Proposition 7. _The integral_ $I(f, e, t, \omega)$ _is_ 0 _unless the conductor of_ ω _divides_ $\mathfrak{f}$. _It is also given by_

$$(14)\qquad I(f, e, t, \omega) = \int_{k_A^1} c(dftu)\psi(etu)\omega(u)d^\times u \ .$$

If the conductor of ω *is* $\mathfrak{f}$, *we have*

$$I(f, e, t, \omega) = \kappa^{-1}\omega(dft)^{-1}|\mathfrak{f}|^{-1/2}\Big[\prod_{v/\mathfrak{f}}(1-q_v^{-1})\omega_v(e_v)\Big]^{-1}\cdot Z_n(\omega) ,$$

where n *is the degree of* div(dft), *and* κ *is as in* §10.

For any place v, take $s \in r_v^\times$ if $|f|_v = 1$, and otherwise $s \in 1 + f_v r_v$; in the first integral in (12), replace f, e by fs, es, and then u by us^{-1}. On the one hand, this does not change the value of the integral, as we have observed above; on the other hand, it multiplies it with $\omega_v(s)^{-1}$; therefore the integral is 0 unless ω_v is trivial for all s chosen as above. As this is so for all v, our first assertion is proved. Now we will show that the integrand in (14) has compact support on k_A^1. In fact, if we put $\mathfrak{m}$ = div(dftu), this must be positive for u in that support, since otherwise $c(\mathfrak{m})$ would be 0. As $|u| = 1$, div(u) is of degree 0, so that $\deg(\mathfrak{m})$ must be equal to the degree n of div(dft), and $\mathfrak{m}$ itself must be one of the finitely many positive divisors of that degree. As the morphism u $\longrightarrow$ div(u) of $k_A^\times$ into the group $\mathfrak{M}$ of all divisors has the compact kernel $\prod r_v^\times$, this proves our assertion about the integrand in (14). From this, it follows at once that (14) can be rewritten as the second integral in (12). Now, for each positive divisor $\mathfrak{m}$ of degree n, choose an idele m such that $\mathfrak{m}$ = div(m); in view of what has just been said, we have

$$I(f, e, t, \omega) = \omega(dft)^{-1}\sum_{\mathfrak{m}} c(\mathfrak{m})\omega(m)\int\psi(ed^{-1}f^{-1}mu)\omega(u)d^\times u ,$$

where the sum is taken over all positive divisors $\mathfrak{m}$ of degree n, and the integrals are taken over the group $\prod r_v^\times$. Now each one of these integrals splits into a product of local integrals, which can be evaluated by means of the results recalled in Chapter II, §10; in particular, if the conductor of ω is $\mathfrak{f}$, those results show that the integral in question is 0 unless, for each v occurring in $\mathfrak{f}$, the order of m_v is 0, i.e. unless $\mathfrak{m}$ is disjoint from $\mathfrak{f}$. Then the same results give at once for $I(f, e, t, \omega)$ the value stated in our proposition. We have assumed here that $d^{\times}u$ has been normalized so that the group $\prod r_v^{\times}$ has the measure 1.

27. It is usual to say that a function Φ on G_A, left-invariant under G_k, is "cuspidal" (or "parabolic") if it satisfies the condition

$$\int_{k_A/k} \Phi((1, y)g)dy = 0$$

for all $g \in G_A$. With our usual notations, this implies clearly that $c_o = c'_o = 0$, as we see by taking $g = (x, 0)$ and $g = (x, 0)\mathfrak{d}$.

We will say that Φ is B-<u>cuspidal</u> if it merely satisfies the condition

$$\int_{k_A/k} \Phi((x, y))dy = 0$$

for all $x \in k_A^{\times}$, i.e. if the "constant term" $c_o(x)$ in the Fourier series for $F(x, y)$ is 0. We can now state our first main result:

Theorem 2. *Let k be a function-field; let Φ, Φ' be an automorphic B-cuspidal pair on G_A; let $Z(\omega)$, $Z'(\omega)$ be the extended Dirichlet series attached respectively to Φ and to Φ'; assume that they are absolutely convergent somewhere. Then, if ω has a conductor $\mathfrak{f} = \mathrm{div}(f)$ disjoint from $\mathfrak{a}$, the partial sums $Z_n(\omega)$, $Z'_n(\omega)$ are 0 except for $0 \leqq n \leqq \deg[\mathrm{div}(ad^2f^2)]$; the sums $\Sigma Z_n(\omega)$, $\Sigma Z'_n(\omega)$ define a holomorphic continuation of $Z(\omega)$, $Z'(\omega)$ over the whole group of such quasicharacters ω, and, over that group, they satisfy the functional equation*

$$Z(\omega) = a(\mathfrak{f})\omega(\mathfrak{a})\varepsilon(\omega)^2 Z'(\omega^{-1}) , \tag{15}$$

where $\varepsilon(\omega) = \kappa\omega(df)$ is the constant factor in the functional equation for the L-series attached to ω.

In fact, for a given ω, take the idele f in (13) such that the divisor $\mathfrak{f} = \mathrm{div}(f)$ is the conductor of ω; then (13), combined with the conclusion of proposition 7, gives at once:

$$(16) \qquad Z_n(\omega) = \alpha(\mathfrak{f})\omega(\mathfrak{a})\varepsilon(\omega)^2 Z'_m(\omega^{-1}) ,$$

n and m being the degrees of div(dft) and $\mathrm{div}(dfat^{-1})$, respectively; it is merely necessary to observe that, because of the relation between e and e' in proposition 4 of §17, we have $\omega_v(-a_v e_v e'_v) = 1$ for all places v in $\mathfrak{f}$, and also that the product $\prod \omega_v(-1)$, taken over the places v in the conductor $\mathfrak{f}$ of ω, is equal to $\omega(-1)$ and therefore to 1. Now we have $Z_n = 0$ for $n < 0$, $Z'_m = 0$ for $m < 0$, and $n + m$ is the degree of $\mathrm{div}(ad^2f^2)$. Our conclusions follow from this at once.

Now let assumptions be as in theorem 2, and let T_v be the Hecke operator belonging to a place v outside $\mathfrak{a}$; then proposition 5 of Chapter VI, §23, shows that $T_v\Phi$, $T_v\Phi'$ are also B-cuspidal, so that, in view of the results of Chapter VI, §25, the functions $T_v\Phi$, $\alpha(\mathfrak{p})T_v\Phi'$ make up an automorphic pair of B-cuspidal functions, so that we can apply theorem 2 to the corresponding extended Dirichlet series; these series can be immediately written, in terms of $Z(\omega)$, $Z'(\omega)$, by applying proposition 5 of Chapter VI, §23. More generally, if $\mathfrak{n}$ is any divisor disjoint from $\mathfrak{a}$, we can apply theorem 2 to the pair $T_{\mathfrak{n}}\Phi$, $\alpha(\mathfrak{n})T_{\mathfrak{n}}\Phi'$, with similar results.

28. When we drop the assumption that Φ, Φ' are B-cuspidal, theorem 2 must be replaced by the following:

Proposition 8. _Let_ Φ, Φ' _be an automorphic pair on_ G_A; _let_ $Z(\omega)$, $Z'(\omega)$ _be the corresponding extended Dirichlet series. Then the conclusions of theorem 2 are still valid except on the group of quasi-characters_ ω _of_ $k_A^\times/k^\times$ _with the conductor_ 1. _Assume now that there is a place_ v _outside_ $\mathfrak{a}$ _such that_ Φ, Φ' _are eigenfunctions of_ T_v; _put_ $T_v\Phi = \lambda\Phi$; _call_ $\mathfrak{p}$ _the prime divisor corresponding to_ v; _put_

$q_v = |\varphi|^{-1}$, and

$$P(\omega) = [q_v - \lambda\omega(\varphi) + a(\varphi)\omega(\varphi)^2].[1 - \lambda\omega(\varphi) + q_v a(\varphi)\omega(\varphi)^2] ,$$

where ω is any quasicharacter of $k_A^\times/k^\times$ of conductor 1. Then the conclusions of theorem 2 are valid also on the group of such quasi-characters, except that $Z(\omega)$, $Z'(\omega)$ are meromorphic there, but not necessarily holomorphic, while $P(\omega)Z(\omega)$, $P(\omega^{-1})Z'(\omega)$ are holomorphic.

Making at first no assumption on Φ, Φ' except that they are an automorphic pair, we note that, if we replace u by $u_o u$ in the integral in (13), with $\operatorname{div}(u_o) = 1$, i.e. $u_o \in \prod r_v^\times$, the integrand is multiplied with $\omega(u_o)$; therefore the integral is 0 unless ω is trivial on all $r_v^\times$, i.e. unless its conductor is 1. Consequently, the calculations of §27 remain valid whenever this is not so.

The group $k_A^1/k^\times \prod r_v^\times$ is no other than the group of divisor-classes of degree 0 for k; it is finite, and has finitely many characters. In other words, the group of quasicharacters of $k_A^\times/k^\times$ has only finitely many components belonging to quasicharacters of conductor 1. If ω is one of these, the same calculations as in §27 (or indeed a much simpler one) gives

$$\text{(17)}\qquad \begin{aligned} &\omega(t)^{-1}[Z_n(\omega)\omega(d)^{-1} - Z'_m(\omega^{-1})\omega(da)] \\ &\qquad = \int_{k_A^1/k^\times} [c'_o(at^{-1}u^{-1}) - c_o(tu)]\omega(u)d^\times u , \end{aligned}$$

where n, m are respectively the degrees of $\operatorname{div}(dt)$ and $\operatorname{div}(dat^{-1})$; it is easily verified that both sides are unchanged if ω is replaced by any quasicharacter $\omega_s\omega$ in the same connected component of Ω_k; they are both multiplied with $\omega(t_1)$ if t is replaced by tt_1 with $|t_1| = 1$.

No conclusion can be derived from this if one makes no additional assumption about the pair Φ, Φ'. Now assume $T_v\Phi = \lambda\Phi$; then, according to Chapter VI, §25, $T_v \Phi' = \lambda \alpha(\psi)\Phi'$. We can apply proposition 6 of Chapter VI, §24, but it is more convenient to make use of the difference equation for c_o, c'_o, given by proposition 5 of the same Chapter; this shows that, for all x, $\nu \longrightarrow c_o(x\pi_v^\nu)$ and $\nu \longrightarrow c'_o(x\pi_v^\nu)$ are respectively solutions of the difference equations

$$q_v f(\nu+1) - \lambda f(\nu) + \alpha(\psi) f(\nu-1) = 0 \ ,$$
$$q_v f'(\nu+1) - \lambda \alpha(\psi)^{-1} f'(\nu) + \alpha(\psi)^{-1} f'(\nu-1) = 0 \ .$$

Let ρ_1, ρ_2 be the two roots of $q_v T^2 - \lambda T + \alpha(\psi) = 0$; then every solution of the first difference equation is a linear combination of the two fundamental solutions $\nu \longrightarrow \rho_1^\nu$, $\nu \longrightarrow \rho_2^\nu$ if $\rho_1 \neq \rho_2$, and $\nu \longrightarrow \rho_1^\nu$, $\nu \longrightarrow \nu\rho_1^{\nu-1}$ if $\rho_1 = \rho_2$. Similarly, the fundamental solutions of the second equation are $\nu \longrightarrow \rho_1'^{\nu}$, $\nu \longrightarrow \rho_2'^{\nu}$ in the former case, and $\nu \longrightarrow \rho_1'^{\nu}$, $\nu \longrightarrow \nu\rho_1'^{\nu-1}$ in the latter case, with $\rho'_1 = q_v^{-1}\rho_1^{-1}$, $\rho'_2 = q_v^{-1}\rho_2^{-1}$. To simplify notations, we will assume $\rho_1 \neq \rho_2$, since the calculations are similar, and the final conclusions are the same, in the case $\rho_1 = \rho_2$. We may therefore write uniquely $c_o(x\pi_v^\nu)$ in the form $a_1(x)\rho_1^\nu + a_2(x)\rho_2^\nu$; replacing x by $x\pi_v$, we see that $a_i(x\pi_v) = a_i(x)\rho_i$ for $i = 1, 2$. Similarly we can write $c'_o(x)$ as $b_1(x) + b_2(x)$ with $b_i(x\pi_v) = b_i(x)\rho'_i$ for $i = 1, 2$. Substituting these for c_o, c'_o in (17), we get

$$\omega(d)^{-1}[Z_n(\omega) + \sum_{i=1,2} A_i(t, \omega)] = \omega(da)[Z'_m(\omega^{-1}) + \sum_{i=1,2} B_i(t, \omega^{-1})]$$

where we have put

$$A_i(t, \omega) = \omega(dt) \int a_i(tu)\omega(u)d^\times u \ ,$$
$$B_i(t, \omega) = \omega(dat^{-1}) \int b_i(at^{-1}u)\omega(u)d^\times u \ .$$

This shows that $A_i(t, \omega)$, $B_i(t, \omega)$ do not change when t is replaced by tu_o with $|u_o| = 1$, or in other words that they depend only upon the degree of div(t), or, what amounts to the same, only upon n, or upon m (since $n + m$ is the degree D of ad^2); we may therefore write $A_{n,i}(\omega)$ instead of $A_i(t, \omega)$, and $B_{m,i}(\omega)$ instead of $B_i(t, \omega)$. Also, we have $A_i(t\pi_v, \omega) = A_i(t, \omega)\rho_i\omega(\mathfrak{y})$, and a similar relation for B_i; putting now $\delta = \deg(\mathfrak{y})$, we get

$$A_{n+\delta, i}(\omega) = A_{n,i}(\omega)\rho_i\omega(\mathfrak{y}), \quad B_{m+\delta, i}(\omega) = B_{m,i}(\omega)\rho_i'\omega(\mathfrak{y}) \; .$$

Now, always with $m+n = D$, and putting $\delta_n = 1$ for $n \geqq 0$, $\delta_n = 0$ for $n < 0$, put

$$\tilde{Z}_n(\omega) = Z_n(\omega) + \delta_n \sum_{i=1,2} [A_{n,i}(\omega) - \omega(ad^2)B_{m,i}(\omega^{-1})] \; ,$$

$$\tilde{Z}'_m(\omega) = Z'_m(\omega) + (1 - \delta_n) \sum_{i=1,2} [B_{m,i}(\omega) - \omega(ad^2)A_{n,i}(\omega^{-1})] \; .$$

Then we have

$$\tilde{Z}_n(\omega) = \omega(ad^2)\tilde{Z}'_m(\omega) \; ,$$

with $\tilde{Z}_n(\omega) = 0$ for $n < 0$, and $\tilde{Z}'_m(\omega) = 0$ for $m < 0$ and $n \geqq 0$, i.e. for $n \geqq (D + 1)^+$; therefore only finitely many $\tilde{Z}_n$, $\tilde{Z}'_m$ are $\neq 0$ and the sums $\sum_n \tilde{Z}_n(\omega)$, $\sum_m \tilde{Z}'_m(\omega)$ are everywhere defined and holomorphic on the group Ω_k^1 of quasicharacters of conductor 1. Now consider for instance the sum

$$\sum_n \delta_n A_{n,i}(\omega) = \sum_{j=0}^{\delta-1} \sum_{h=0}^{+\infty} A_{j+\delta h, i}(\omega)$$
$$= \sum_{j=0}^{\delta-1} A_{j,i}(\omega) \sum_{h=0}^{+\infty} \rho_i^h \omega(\mathfrak{y})^h \; .$$

If we put $|\omega| = \omega_\sigma$, this is convergent for σ large enough, and is meromorphic on Ω_k^1, its poles being the points for which $\omega(\mathfrak{y}) = \rho_i^{-1}$. Its analytic continuation is of course given by

$$\sum_j A_{j,i}(\omega).[1 - \rho_i\omega(\mathfrak{y})]^{-1} \quad ,$$

so that, for $\sigma < -\sigma_1$ with σ_1 large enough, it can be written

$$-\sum_j A_{j,i}(\omega) \sum_{h=1}^{\infty} \rho_i^{-h}\omega(\mathfrak{y})^{-h} = -\sum_n (1-\delta_n)A_{n,i}(\omega) \quad .$$

Treating $B_{m,i}$ similarly, we get the conclusion of our proposition, at least for $\rho_1 \neq \rho_2$. As we have said, we omit the calculations for the case $\rho_1 = \rho_2$, which are quite similar.

Remark. The definition of $A_{n,i}(\omega)$, $B_{m,i}(\omega)$ shows that, essentially, they are the Fourier coefficients of $a_i(x)$, $b_i(x)$ on the cosets of the compact group $k_A^1/k^\times$ in the group $k_A^\times/k^\times$, so that they determine the a_i, b_i, and consequently c_o, c'_o uniquely; this could easily be expressed by simple formulas. At the same time, the above calculations show that $A_{n,i}(\omega)$, $B_{m,i}(\omega)$ are uniquely determined by the "principal part" of $Z(\omega)$ at its poles; for $\rho_1 \neq \rho_2$, for instance, they can easily be expressed in terms of its residues. Consequently, those principal parts determine c_o, c'_o uniquely (and explicit formulas can easily be written to express this dependence).

29. We shall now be concerned with the converse of theorem 2; in doing this, we will simplify our calculations by confining ourselves to the case of a B-cuspidal pair Φ, Φ', which is in various respects the most interesting one to consider.

Consequently, we start with two functions F, F' on B_A, given by their Fourier series with the coefficients $c(\mathfrak{m})$, $c'(\mathfrak{m})$, the "constant terms" c_o, c'_o being both 0. We write $Z(\omega)$, $Z'(\omega)$ for the extended Dirichlet series with these coefficients; we assume that these series are convergent somewhere, i.e. that $c(\mathfrak{m})$, $c'(\mathfrak{m})$ are $O(|\mathfrak{m}|^{-\alpha})$ for some α. In order to obtain sufficient conditions for F, F' to be induced on B_A by an automorphic pair Φ, Φ', we apply theorem 1 of Chapter IV, §18, and therefore choose a set S of places of k, including all the places in $\mathfrak{a}$, with the "approximation property" (as defined there); in particular, we may take for S any finite set. Then, by theorem 1, a sufficient condition is given by (II), or, what amounts to the same, by (II'); in view of the calculations in §26, this is expressed by (10), or, since $c_o = c'_o = 0$, by

$$(18) \qquad I(f, e, t, \omega) = a(\mathfrak{f})I'(f, e', at^{-1}, \omega^{-1}) ,$$

where the left-hand side is given by (14), the right-hand side by the similar formula with c' substituted for c, and where $\mathfrak{f}$ is disjoint from S. By proposition 7 of §26, this is trivially fulfilled when the conductor of ω does not divide $\mathfrak{f}$, since then both sides of (18) are 0. If the conductor of ω is $\mathfrak{f}$, the value of both sides is given by proposition 7, and the calculations in the proof of theorem 2, §27, show that (18) is equivalent to (16) and implies (15). Conversely, assume that $Z(\omega)$, $Z'(\omega^{-1})$ can be continued as holomorphic functions over some connected component of the group Ω_k and satisfy (15) there; replacing ω by $\omega_s\omega$, and taking in both sides the part which is homogeneous of degree n in ω_s, we get (16).

30. Let $\mathfrak{n}$ be a positive divisor, disjoint from $\mathfrak{a}$, so that the Hecke operator $T_{\mathfrak{n}}$ is defined; we will write $I_{\mathfrak{n}}(f, e, t, \omega)$ for the integral, similar to (12), but where $T_{\mathfrak{n}}F$ has been substituted for F. If we take for $\mathfrak{n}$ the prime divisor $\mathfrak{p}$ belonging to a place v outside $\mathfrak{a}$, we can use formula (9) of Chapter VI, §23, to express T_vF; at the same time, having assumed that Φ is B-cuspidal, we know that also $T_v\Phi$ is so; therefore, in (12), there is no term in c_o, either for F or for T_vF. We get now from (12):

$$I_{\mathfrak{p}}(f, e, t, \omega) = \sum_r \int_{k_A^1/k^\times} F(tuf\pi_v, tu(e + fr))\omega(u)d^\times u$$
$$+ a(\mathfrak{p}) \int_{k_A^1/k^\times} F(tuf\pi_v^{-1}, tue)\omega(u)d^\times u ,$$

where the sum is taken over a complete set of representatives of the classes modulo π_v in r_v.

Assume first that $\mathfrak{p}$ divides $\mathfrak{f}$; then $|e|_v = 1$; substituting then $u(1 + e_v^{-1} f_v r)^{-1}$ for u in the integral corresponding to r, and observing that $(f\pi_v, e)$ and $(f\pi_v^{-1}, e)$ are both reduced, we get

$$I_{\mathfrak{p}}(f, e, t, \omega) = I(f\pi_v, e, t, \omega) \sum_r \omega_v(1 + e_v^{-1} f_v r)^{-1} + \alpha(\mathfrak{p}) I(f\pi_v^{-1}, e, t, \omega) .$$

Here the sum in the right-hand side is q_v or 0 according as ω_v is trivial on $1 + f_v r_v$ or not; if we assume that the conductor of ω divides $\mathfrak{f}$ (which is the only interesting case, since otherwise both sides in the above formula are 0), we get

$$I_{\mathfrak{p}}(f, e, t, \omega) = q_v I(f\pi_v, e, t, \omega) + \alpha(\mathfrak{p}) I(f\pi_v^{-1}, e, t, \omega) .$$

Now take the case where $\mathfrak{p}$ does not divide $\mathfrak{f}$. Then $(f, e\pi_v)$ is reduced, so that the last integral in the formula for $I_{\mathfrak{p}}$ is $I(f, e\pi_v, t\pi_v^{-1}, \omega)$, which we may rewrite as $I(f, e, t\pi_v^{-1}, \omega)$, since we know that the latter integral does not depend upon the value of e_v. As to the other integrals, put $s = e_v + f_v r$, and observe that, for $s \equiv 0 \bmod. \pi_v$, $(f, (e + fr)\pi_v^{-1})$ is reduced; the corresponding integral can therefore be rewritten as $I(f, e, t\pi_v, \omega)$, since this does not depend upon the value of e_v. For $s \in r_v^\times$, $(f\pi_v, e + fr)$ is reduced; if, in the integral, we replace u by us^{-1}, and make use of the known properties of I, we find for the integral the value $\omega_v(s)^{-1} I(f\pi_v, e_1, t, \omega)$, where e_1 has at every place of k the same component as e, except at v where it has the component 1. Here s goes through a set of representatives of the classes modulo $1 + \pi_v r_v$ in $r_v^\times$, so that $\Sigma\omega_v(s)^{-1}$ is 0 or $q_v - 1$ according as $\mathfrak{p}$ occurs in the conductor of ω or not; moreover, in the latter case, this shows that the integral is independent of the v-component of e_1, so that we may rewrite it as $I(f\pi_v, e, t, \omega)$ provided $|e|_v = 1$. Here again the significant case for us is the one where the conductor of ω divides $\mathfrak{f}$; then $\mathfrak{p}$ does not occur in it, and we get

$$I_{\mathfrak{p}}(f, e, t, \omega) = (q_v - 1)I(f\pi_v, e, t, \omega) + I(f, e, t\pi_v, \omega) + \alpha(\mathfrak{p})I(f, e, t\pi_v^{-1}, \omega) .$$

Now, in the formulas we have just found, substitute $f\pi_v^{-1}$ for f; we get

$$(19) \qquad q_v I(f, e, t, \omega) = I_{\mathfrak{p}}(f\pi_v^{-1}, e, t, \omega) - \alpha(\mathfrak{p})I(f\pi_v^{-2}, e, t, \omega) ,$$

$$(20) \qquad (q_v - 1)I(f, e, t, \omega) = I_{\mathfrak{p}}(f\pi_v^{-1}, e, t, \omega) - I(f\pi_v^{-1}, e, t\pi_v, \omega) + \alpha(\mathfrak{p})I(f\pi_v^{-1}, e, t\pi_v^{-1}, \omega) ;$$

in both formulas, it is assumed that $\mathfrak{p}$ divides $\mathfrak{f}$, and that the conductor of ω divides $\mathfrak{f}\mathfrak{p}^{-1}$; the first one is valid if $\mathfrak{p}^2$ divides $\mathfrak{f}$, and the second one if that is not so.

31. Always with the same notations, we are now ready to prove the decisive lemma:

Lemma 5. _Take_ f_o _so that_ $\mathfrak{f}_o = \mathrm{div}(f_o)$ _is the conductor of_ ω, _and assume that it divides_ $\mathfrak{f}$. _Assume that we have_

$$I_{\mathfrak{n}}(f_o, e, t, \omega) = \alpha(\mathfrak{n}\mathfrak{f}_o)I'_{\mathfrak{n}}(f_o, e', at^{-1}, \omega^{-1})$$

for all positive divisors $\mathfrak{n}$ _dividing_ $\mathfrak{f}_o^{-1}\mathfrak{f}$, _and all_ e, e', t (_where_ e, e' _satisfy the same conditions as before_). _Then we have_

$$I(f, e, t, \omega) = \alpha(\mathfrak{f})I'(f, e', at^{-1}, \omega^{-1}) .$$

Take f' so that $\mathfrak{f}' = \mathrm{div}(f')$ is a multiple of $\mathfrak{f}_o$ and divides $\mathfrak{f}$; by induction on the number of prime divisors of $\mathfrak{f}_o^{-1}\mathfrak{f}'$ (distinct or not) we will prove that the assumption in our lemma implies that we have

$$I_{\mathfrak{n}}(f', e, t, \omega) = \alpha(\mathfrak{n}\mathfrak{f}')I'_{\mathfrak{n}}(f', e', at^{-1}, \omega^{-1})$$

provided $\mathfrak{n}$ divides $\mathfrak{f}'^{-1}\mathfrak{f}$; for $\mathfrak{f} = \mathfrak{f}'$, this will prove our lemma. Write $[\mathfrak{f}', \mathfrak{n}]$ for the set of relations to be proved by induction; taking for $\mathfrak{y}$ a prime divisor of $\mathfrak{f}_o^{-1}\mathfrak{f}'$, the induction assumption says that $[\mathfrak{f}'\mathfrak{y}^{-1}, \mathfrak{n}\mathfrak{y}]$, $[\mathfrak{f}'\mathfrak{y}^{-1}, \mathfrak{n}]$, and also $[\mathfrak{f}'\mathfrak{y}^{-2}, \mathfrak{n}]$ if $\mathfrak{y}^2$ divides $\mathfrak{f}_o^{-1}\mathfrak{f}'$, are all valid. If $\mathfrak{y}^2$ divides $\mathfrak{f}'$ but not $\mathfrak{f}_o^{-1}\mathfrak{f}'$, then $[\mathfrak{f}'\mathfrak{y}^{-2}, \mathfrak{n}]$ is trivially true, since in that case both sides in it are 0. It will therefore be enough to show that these relations imply $[\mathfrak{f}', \mathfrak{n}]$. Writing now F, F' instead of $T_{\mathfrak{n}}F$, $T_{\mathfrak{n}}F'$, and f instead of f', we see that it is enough for us to show that, if $\mathfrak{y}$ is any prime divisor of $\mathfrak{f}_o^{-1}\mathfrak{f}$, the validity of the relations $[\mathfrak{f}\mathfrak{y}^{-1}, \mathfrak{y}]$, $[\mathfrak{f}\mathfrak{y}^{-1}, 1]$, and $[\mathfrak{f}\mathfrak{y}^{-2}, 1]$ if $\mathfrak{y}^2$ divides $\mathfrak{f}$, implies that of the relation in the conclusion of our lemma. Now, in that conclusion, replace both sides by their values as given by (19) or by (20), as the case may be. A trivial calculation gives our conclusion at once.

32. We can now prove the theorem which concludes our investigation for the function-field case:

Theorem 3. Let S be a set of places of k, including those in $\mathfrak{a}$, with the approximation property; let Ω_S be the group of the quasi-characters of $k_A^\times/k^\times$ whose conductor is disjoint from S. Let $Z(\omega)$, $Z'(\omega)$ be two extended Dirichlet series, both convergent somewhere; let F, F' be the functions defined on B_A by the Fourier series (without constant terms) with the same coefficients as $Z(\omega)$ and as $Z'(\omega)$, respectively. Then F and F' can be extended to an automorphic B-cuspidal pair (Φ, Φ') if and only if, for all positive divisors $\mathfrak{n}$ disjoint from S, the extended Dirichlet series $Z_{\mathfrak{n}}(\omega)$, $Z'_{\mathfrak{n}}(\omega)$, derived from $Z(\omega)$, $Z'(\omega)$ by the Hecke operators $T_{\mathfrak{n}}$, can be continued as holomorphic functions on Ω_S and satisfy on Ω_S the functional equation

$$Z_{\mathfrak{n}}(\omega) = a(\mathfrak{f}\mathfrak{n})\omega(\mathfrak{d})\varepsilon(\omega)^2 Z'_{\mathfrak{n}}(\omega^{-1}) \ ,$$

where notations are as in theorem 2.

To say that the condition is necessary is merely to repeat the statement of theorem 2. As to the converse, we have seen in §29 that we merely have to verify the validity of the conclusion of lemma 5 for all $\mathfrak{f} = \mathrm{div}(f)$ disjoint from S, and all ω whose conductor divides $\mathfrak{f}$ and is therefore also disjoint from S. This will be valid, if the assumption in lemma 5 is satisfied; but then we merely have to make use of proposition 7 of §26, just as in the proof of theorem 2, to complete the proof of our theorem.

Corollary. Assumptions being as in theorem 3, assume also that for every prime divisor $\mathfrak{y}$ not in S, there is λ such that F and F' (or Z and Z') are eigenfunctions of $T_{\mathfrak{y}}$ for the eigenvalues λ, $\lambda a(\mathfrak{y})^{-1}$, respectively. Then theorem 3 remains valid if we restrict the condition in it to the case $\mathfrak{n} = 1$.

This is obvious.

From the results of §§24-25 it follows at once that the condition on Z and Z' in the above corollary is fulfilled if (and only if) they are eulerian at all places $\mathfrak{y}$ not in S, with Euler factors of the form described in §§24-25.

CHAPTER VIII

HARMONICITY AT AN INFINITE PLACE

33. From now on, k will be a number-field.

From the point of view of representation-theory, which is that of Jacquet and Langlands (loc. cit.), the finite and the infinite places do not really play different roles; for both, one studies the infinite-dimensional representations of $G_v = GL(2, k_v)$ and proves the corresponding "local functional equations" before assembling them to obtain automorphic functions and Euler products. Actually, in their theory, the infinite places are in some respects easier to deal with; at any rate, for these places, the representations of G_v have been (in substance) well-known and fully classified for some time, while there is a type of finite places (the "even" ones, i.e. those with the residual characteristic 2) for which this is not yet so.

From the "elementary" point of view adopted here, we need a separate discussion of the infinite places; that is the object of this Chapter; the finite places could presumably also be discussed from a similar point of view, but this is not necessary for our present purposes and will not be attempted. The main results of this Chapter are entirely due to Jacquet and Langlands (even though important special cases were already implicit in the work of Hecke and in that of Maass). It is probably true that a deeper understanding of these results requires a reference to their theory. Nevertheless, the treatment to be given here (based partly on their publication, and partly on Godement's Notes, loc. cit.) will be self-contained; except for occasional side-remarks, representations will not be mentioned. Typical special cases, relevant for the classical theory and for the theory of the zeta-function of elliptic curves, will be described in the next Chapter; the reader interested only in those cases (or in getting a bird's eye view of the theory) may therefore skip this one entirely in a first reading, except for §§33-35.

In this Chapter and the next one, we select once for all one infinite place w of k, write K for the local field k_w, and suppress the subscript w altogether. Thus K is either $\mathbf{R}$ or $\mathbf{C}$, and we write G for $GL(2, K)$, $\mathfrak{z}$ for the center of G, and B for the subgroup of G consisting of the matrices $\begin{pmatrix} x & y \\ 0 & 1 \end{pmatrix}$.

We will say that a $\mathbf{C}$-valued function f on G is <u>B-moderate</u> if there is $N \geqq 0$ such that

$$|f(\begin{pmatrix} p & 0 \\ 0 & 1 \end{pmatrix} . g)| = O(p^N)$$

for $p \in \mathbf{R}$, $p \longrightarrow +\infty$, uniformly over compact sets with respect to g in G. We extend this definition in an obvious manner to functions with values in a finite-dimensional vector-space V over $\mathbf{C}$. For a B-moderate function, we have, more generally:

$$|f(\begin{pmatrix} x & 0 \\ 0 & 1 \end{pmatrix} . g)| = O(|x|^N)$$

for $x \in K^\times$, $|x| \longrightarrow +\infty$, uniformly over compact sets with respect to g.

34. As before, we write $\mathfrak{K}$ for $O(2, \mathbf{R})$ resp. $U(2, \mathbf{C})$ according as $K = \mathbf{R}$ or $K = \mathbf{C}$. We will write G_1 for the subgroup $SL(2, K)$ of G determined by $\det g = 1$, and $\mathfrak{K}_1$ for $\mathfrak{K} \cap G_1$, i.e. $SO(2, \mathbf{R})$ resp. $SU(2, \mathbf{C})$. We will write G^o for the connected component of 1_2 in G; this is G itself, if $K = \mathbf{C}$; if $K = \mathbf{R}$, it is the subgroup of G determined by $\det g > 0$. Then we have $G^o = G_1 \mathfrak{z}$. On the other hand, we will write B_1 for the subgroup of G_1 consisting of the matrices

$$b_1 = p^{-1/2} . \begin{pmatrix} p & y \\ 0 & 1 \end{pmatrix}$$

for $p \in \mathbf{R}$, $p > 0$, $y \in K$. Then we have $G_1 = B_1\mathfrak{K}_1$, $G^o = B_1\mathfrak{K}_1\mathfrak{Z}$, $G = B_1\mathfrak{K}\mathfrak{Z}$. More precisely, B_1 is a complete set of representatives for the cosets $g\mathfrak{K}_1$ in G_1 and $g\mathfrak{K}\mathfrak{Z}$ in G and may be identified with the space $H = G_1/\mathfrak{K}_1 = G/\mathfrak{K}\mathfrak{Z}$, i.e. with the Riemannian symmetric space belonging to G_1 (or to G); it may also, in the obvious manner, be identified with the "half-plane" (for $K = \mathbf{R}$) or the "half-space" (for $K = \mathbf{C}$) consisting of the points (p, y) with $p > 0$ in the plane (resp. in the space) $\mathbf{R} \times K$.

35. As we suppress the subscript w, we write ψ for the character of K given by $x \longrightarrow e^{-2\pi i x}$ if $K = \mathbf{R}$, by $x \longrightarrow e^{-2\pi i(x+\bar{x})}$ if $K = \mathbf{C}$; we write α for the character of $K^\times$ previously denoted by α_w (§12). If $K = \mathbf{R}$, we can write α (uniquely) as

$$\alpha(z) = (\operatorname{sgn} z)^m |z|^\zeta \tag{21}$$

with $m = 0$ or 1, $\zeta \in \mathbf{C}$. If $K = \mathbf{C}$, we can write α (uniquely) as $z \longrightarrow z^m(z\bar{z})^{\zeta''}$, with $m \in \mathbf{Z}$, $\zeta'' \in \mathbf{C}$; then we put $\zeta' = m + \zeta''$ and write more briefly (by "abuse of language")

$$\alpha(z) = z^{\zeta'}\bar{z}^{\zeta''} , \tag{22}$$

with $\zeta' \equiv \zeta''$ mod. 1. If a function Φ on G satisfies $\Phi(g\mathfrak{z}) = \Phi(g)\alpha(\mathfrak{z})$ for all $g \in G$, $\mathfrak{z} \in \mathfrak{Z}$, it is uniquely determined on G^o (if $K = \mathbf{R}$) and on G (if $K = \mathbf{C}$) by its values on G_1.

In Chapter IV, §14, we introduced an irreducible representation M_∞ of $\mathfrak{K}_\infty$, made up of representations M_w of the groups $\mathfrak{K}_w$. Accordingly, here, we introduce an irreducible representation M of $\mathfrak{K}$, of which we assume that it coincides with α on $\mathfrak{K} \cap \mathfrak{Z}$, and we call V its representation-space; as before, we can extend it to $\mathfrak{K}\mathfrak{Z}$ by putting $M(\mathfrak{k}\mathfrak{z}) = M(\mathfrak{k})\alpha(\mathfrak{z})$; we will say that a V-valued function Φ on G <u>belongs to</u> M if $\Phi(g\mathfrak{k}\mathfrak{z}) = \Phi(g)M(\mathfrak{k}\mathfrak{z})$ for all g in G and $\mathfrak{k}\mathfrak{z}$ in $\mathfrak{K}\mathfrak{Z}$; then Φ is uniquely determined by its values on B_1.

The irreducible representations of $\mathfrak{K}_1$, and those of $\mathfrak{K}$, which coincide with α, are easily classified, up to equivalence:

a) For $K = \mathbf{R}$, put

$$r(\theta) = \begin{pmatrix} \cos\theta & \sin\theta \\ -\sin\theta & \cos\theta \end{pmatrix}, \quad s = \begin{pmatrix} -1 & 0 \\ 0 & 1 \end{pmatrix};$$

$\mathfrak{K}_1$ consists of the elements $r(\theta)$, and $\mathfrak{K}$ of the elements $r(\theta)$, $r(\theta)s$. The irreducible representations of $\mathfrak{K}_1$ are those given by $r(\theta) \longrightarrow e^{i\nu\theta}$, with $\nu \in \mathbf{Z}$. Those of $\mathfrak{K}$ are, firstly, those given by

$$M_n(r(\theta)) = \begin{pmatrix} e^{in\theta} & 0 \\ 0 & e^{-in\theta} \end{pmatrix}, \quad M_n(s) = \begin{pmatrix} 0 & 1 \\ 1 & 0 \end{pmatrix}$$

for $n \in \mathbf{Z}$, $n > 0$; in addition to these, we have the trivial representation $M_o = 1$, and another one M'_o, trivial on $\mathfrak{K}_1$, with $M'_o(s) = -1$. As these must coincide on the center $\{\pm 1_2\}$ with the character α given by (21), we must have $\nu \equiv m$ mod. 2 resp. $n \equiv m$ mod. 2.

b) For $K = \mathbf{C}$, put

$$r(\xi, \eta) = \begin{pmatrix} \xi & \eta \\ -\bar{\eta} & \bar{\xi} \end{pmatrix}$$

with $\xi\bar{\xi} + \eta\bar{\eta} = 1$; these make up $\mathfrak{K}_1$. For any integer $n \geqq 0$, take $V = \mathbf{C}^{n+1}$; for $v = (v_o, \ldots, v_n)$ in V, and $r(\xi, \eta)$ as above, define $v' = (v'_o, \ldots, v'_n)$ by the identity

$$\sum_{j=0}^{n} v'_j X^{n-j} Y^j = \sum_{j=0}^{n} v_j (\xi X + \eta Y)^{n-j} (-\bar{\eta} X + \bar{\xi} Y)^j$$

in the two indeterminates X, Y, and write then $v' = v.M_n(r(\xi, \eta))$; this defines an irreducible representation M_n of $\mathfrak{K}_1$ (the trivial one for $n = 0$, the identical one for $n = 1$). It coincides on the center with the character α given by (22) if and only if $\alpha(-1) = (-1)^n$, i.e.

$n \equiv \zeta' - \zeta'' \mod. 2$.

36. For $K = \mathbf{R}$, we will denote by X, Y, U, Z the generators of the Lie algebra of G, respectively given by

$$X = \begin{pmatrix} 0 & 1 \\ 0 & 0 \end{pmatrix}, \quad Y = \begin{pmatrix} 0 & 0 \\ 1 & 0 \end{pmatrix}, \quad U = \begin{pmatrix} 1 & 0 \\ 0 & -1 \end{pmatrix}, \quad Z = \begin{pmatrix} 1 & 0 \\ 0 & 1 \end{pmatrix} .$$

Then X, Y, U generate the Lie algebra of G_1, or, as we may say more briefly (since G_1 is connected), they generate G_1; X and U generate B_1; $X - Y$ generates $\mathfrak{K}_1$; Z generates the Lie algebra and the connected component of $\mathfrak{Z}$. We may regard X, Y, U, Z as left-invariant differential operators on G, given as such e.g. by

$$Xf(g) = \left[\frac{d}{d\tau} f(g.(1_2 + X\tau)) \right]_{\tau=0} .$$

The differential operators which are invariant under both right-translations and left-translations are those given by the center of the enveloping algebra of the Lie algebra. On G_1, this center is generated by the so-called Casimir operator

$$D = \frac{U^2}{2} + XY + YX \quad ;$$

on G, it is generated by D and Z.

For $K = \mathbf{C}$, we will consider, for G and its subgroups, the "complex" Lie algebras, i.e. the complexifications of the real Lie algebras. That of G is then the direct product of two algebras, both isomorphic to the matrix algebra $M_2(\mathbf{C})$; it has a basis consisting of elements $X', X'', Y', Y'', U', U'', Z', Z''$, which, when regarded as left-invariant operators, are respectively defined e.g. by

$$X'f(g) = \left[\frac{\partial}{\partial\tau} f(g.(1_2 + X\tau))\right]_{\tau=0} ,$$

$$X''f(g) = \left[\frac{\partial}{\partial\bar{\tau}} f(g.(1_2 + X\tau))\right]_{\tau=0} ,$$

where τ is a complex variable; as usual, if $\tau = u + iv$, $\partial/\partial\tau$ and $\partial/\partial\bar{\tau}$ are respectively defined as $\frac{1}{2}(\frac{\partial}{\partial u} - i\frac{\partial}{\partial v})$ and $\frac{1}{2}(\frac{\partial}{\partial u} + i\frac{\partial}{\partial v})$. Then Z', Z'' "generate" $\mathfrak{z}$; X', X'', $U' + U''$ generate B_1; $X'' - Y'$, $X' - Y''$, $U' - U''$ generate $\mathfrak{K}_1$; X', X'', Y', Y'', U', U'' generate G_1. There are two "Casimir operators"

$$D' = \frac{U'^2}{2} + X'Y' + Y'X', \quad D'' = \frac{U''^2}{2} + X''Y'' + Y''X'' \;;$$

they generate the center of the enveloping algebra for G_1, and, together with Z' and Z'', they generate that of G.

37. If Φ belongs to an irreducible representation of $\mathfrak{K}\mathfrak{z}$, so does $D\Phi$ if $K = \mathbf{R}$ (resp. $D'\Phi$, $D''\Phi$ if $K = \mathbf{C}$); this follows at once from the fact that D (resp. D', D'') is right-invariant as well as left-invariant. As Φ is then determined by the function it induces on B_1, we may thus, for a given M, regard D (resp. D', D'') as operating on functions on B_1. To make this more explicit, we proceed as follows.

For $K = \mathbf{R}$, take φ such that $\varphi(g.r(\theta)) = \varphi(g)e^{i\nu\theta}$; then we have $(X - Y)\varphi = i\nu\varphi$ (and conversely, this reflecting the fact that $X - Y$ generates $\mathfrak{K}_1$). As $XY - YX = U$, this gives

$$(XY + YX)\varphi = 2XY\varphi - U\varphi = 2X(X - i\nu)\varphi - U\varphi .$$

Call f, i.e. $(p, y) \longrightarrow f(p, y)$, the function induced by φ on B_1; as X and U are in the Lie algebra of B_1, an easy calculation shows now that the function f' induced by $D\varphi$ on B_1 is given by

$$f'(p, y) = 2p^2 \left(\frac{\partial^2 f}{\partial p^2} + \frac{\partial^2 f}{\partial y^2} - \frac{i\nu}{p}\frac{\partial f}{\partial y}\right) . \tag{23}$$

The elliptic operator $f \longrightarrow f'$, in the upper half-plane, may be regarded as a generalized Beltrami operator; it is the usual Beltrami operator if $\nu = 0$.

For $K = \mathbb{C}$, let Φ be a function on G, with values in $V = \mathbb{C}^{n+1}$, belonging to the representation M_n of $\mathfrak{K}_1$. Let f, f', f'' be the functions induced on B_1 by Φ, $D'\Phi$, $D''\Phi$, respectively. Put $A_n = M_n(U' - U'')$, $A'_n = M_n(X'' - Y')$, $A''_n = M_n(X' - Y'')$, where M_n has been extended to the Lie algebra of $\mathfrak{K}_1$ in the usual manner; in particular, A_n is a diagonal matrix with the diagonal elements $n - 2j$ $(0 \leqq j \leqq n)$. A calculation, similar to the one outlined above for $K = \mathbb{R}$, gives now

$$f'(p, y) = \frac{1}{2}p^2\left(\frac{\partial^2 f}{\partial p^2} + 4\frac{\partial^2 f}{\partial y \partial \bar{y}}\right) + \frac{1}{2}p\frac{\partial f}{\partial p}\cdot(A_n - 1_{n+1}) \tag{24}$$
$$- 2p\frac{\partial f}{\partial y}\cdot A'_n + \frac{1}{8}f\cdot(A_n^2 - 4A_n) ,$$

and a similar formula for f'' with y, $\bar{y}$, A_n, A'_n replaced by $\bar{y}$, y, $-A_n$, A''_n, respectively. For $n = 0$, we have $M_o = 1$, $A_o = A'_o = A''_o = 0$, $f' = f''$, and $f \longrightarrow f'$ is again the usual Beltrami operator in the upper half-space. For $n > 0$, we may regard $f \longrightarrow f'$ and $f \longrightarrow f''$ as generalized Beltrami operators.

38. Clearly, if a function Φ on G belongs to α, i.e. if $\Phi(g\mathfrak{z}) = \Phi(g)\alpha(\mathfrak{z})$ for all $g \in G$, $\mathfrak{z} \in \mathfrak{Z}$, it is an eigenfunction of Z (resp. of Z', Z'') for the eigenvalue ζ (resp. ζ', ζ''). The converse is also true in the case $K = \mathbb{C}$, as one sees at once (because then $\mathfrak{Z}$ is connected, hence "generated" by Z', Z''). For $K = \mathbb{R}$, $Z\Phi = \zeta\Phi$ only implies that $\Phi(g\mathfrak{z}) = \Phi(g)\alpha(\mathfrak{z})$ for $\mathfrak{z} = z.1_2$, $z > 0$; if at the same time (as will usually be the case) Φ belongs to a given irreducible representation of $\mathfrak{K}_1$ or to one of $\mathfrak{K}$, then (since $-1_2 = r(\pi)$) its behavior under $\mathfrak{z} = -1_2$ is also prescribed, and α is again uniquely

determined by the eigenvalue ζ and that condition.

The $\mathbf{C}$-valued functions on G which belong to a given character α of $\mathfrak{Z}$ and are eigenfunctions of D (resp. D', D'') with prescribed eigenvalues make up a space which is invariant under right- and left-translations; this accounts for the role they play in representation-theory. Under broad conditions, the convolutions of such functions with any function (or distribution) are again of the same type.

A function Φ, with values in $\mathbf{C}$ or in a finite-dimensional vector-space V over $\mathbf{C}$, will be said to be harmonic of type (α, δ) resp. $(\alpha, \delta', \delta'')$ if $\Phi(g\mathfrak{z}) = \Phi(g)\alpha(\mathfrak{z})$ for all $g \in G$, $\mathfrak{z} \in \mathfrak{Z}$, and $D\Phi = \delta\Phi$ (resp. $D'\Phi = \delta'\Phi$, $D''\Phi = \delta''\Phi$). If at the same time it belongs to a given irreducible representation M of $\mathfrak{K}_1$ or of $\mathfrak{K}$, then it will be said to be harmonic of type (α, δ, M) resp. $(\alpha, \delta', \delta'', M)$; more precisely, it will be said to be harmonic of type (α, δ, ν) if $K = \mathbf{R}$ and M is the representation $r(\theta) \longrightarrow e^{i\nu\theta}$ of $\mathfrak{K}_1$, and of type $(\alpha, \delta', \delta'', n)$ if $K = \mathbf{C}$ and M is the representation M_n of $\mathfrak{K}_1$ defined in §35, b). Clearly such functions make up a space which is invariant under left-translations. Moreover, such a function Φ is uniquely determined on G^o by its values on B_1; for $K = \mathbf{R}$, it is uniquely determined on G by the values induced on B_1 by Φ and by $g \longrightarrow \Phi(sg)$, with s as in §35, a). Now (23) resp. (24) shows that the functions which are so induced on B_1 are eigenfunctions of at least one elliptic operator, and are therefore real-analytic; consequently the same is true of Φ. More generally, the same is true of any eigenfunction of the "central" operators $(Z, D$, resp. $Z', Z'', D', D'')$ which are "$\mathfrak{K}_1$-finite," i.e. whose right translates by elements of $\mathfrak{K}_1$ all lie in a finite-dimensional vector-space over $\mathbf{C}$; in fact, one can express such a function as a finite linear combination of components of functions belonging to irreducible representations of $\mathfrak{K}_1$; these are again eigenfunctions of the central operators, and one can apply to them the foregoing remarks.

39. Let φ be harmonic of type (α, δ) resp. $(\alpha, \delta', \delta'')$; let λ be any quasicharacter of $K^\times$; then one sees at once that the function $g \longrightarrow \varphi(g)\lambda(\det g)$ is harmonic of type $(\alpha\lambda^2, \delta)$ resp. $(\alpha\lambda^2, \delta', \delta'')$.

We will now describe some examples of harmonic functions (in a suitable sense, they may be looked upon as "elementary solutions" of the equations for harmonic functions).

a) Let $f(a, c)$ be such that

$$f(ax, cx) = f(a, c)\alpha(x)$$

for all $(a, c) \in K^2 - \{0\}$ and all $x \in K^\times$, and let φ be given by $\varphi(g) = f(a, c)$ for $g = \begin{pmatrix} a & b \\ c & d \end{pmatrix}$. Then one verifies at once that φ is harmonic of type $(\alpha, \frac{1}{2}\zeta^2 + \zeta)$ resp. $(\alpha, \frac{1}{2}\zeta'^2 + \zeta', \frac{1}{2}\zeta''^2 + \zeta'')$. A particularly interesting case (because of the role it will play as an "elementary solution") is that of the function given, for $\xi \in K^\times$, by $g \longrightarrow \varphi(g) = \alpha(c)\psi(\xi\frac{a}{c})$; it is singular for $c = 0$, but its convolutions $\varphi * \mu$ need not be so. Its value to us lies in its behavior under left-translations by elements $(1, y)$:

$$\varphi(\begin{pmatrix} 1 & y \\ 0 & 1 \end{pmatrix} \cdot g) = \psi(\xi y)\varphi(g) \quad ,$$

because of which we shall be able to use it as a generating function for the "Whittaker functions."

b) For $K = \mathbf{R}$, consider the harmonic functions of type (α, δ, n) on G^o which are invariant under left-translations by elements $(1, y)$. Such a function will induce on B_1 a function $(p, y) \longrightarrow f(p)$, where f, in view of (23), must satisfy $2p^2 \frac{d^2f}{dp^2} = \delta f$. This has two linearly independent solutions, given by $f(p) = p^{(1 \pm \rho)/2}$, with $\rho = (1 + 2\delta)^{1/2}$, if $\rho \neq 0$, and by $f(p) = p^{1/2}$, $f(p) = p^{1/2} \log \frac{1}{p}$, if $\rho = 0$.

c) Similarly, for $K = \mathbb{C}$, consider the harmonic functions of type $(\alpha, \delta', \delta'', n)$, left-invariant under $(1, y)$ for all y. As in b), the function $f = (f_o, \ldots, f_n)$ induced by such a function on B_1 must be of the form $(p, y) \longrightarrow f(p)$ and satisfy the differential equation

$$(E'_o) \qquad p^2 \frac{d^2f}{dp^2} + p\frac{df}{dp}\cdot(A_n - 1_{n+1}) + f.[(A_n/2)^2 - A_n - 2\delta'.1_{n+1}] = 0$$

derived from (24), and the similar equation (E''_o) obtained by replacing A_n, δ' by $-A_n$, δ'' in (E'_o). An elementary calculation, the details of which we omit, shows that there is no solution $\neq 0$ unless δ', δ'' are of the form $\delta' = \frac{1}{2}(\rho'^2 - 1)$, $\delta'' = \frac{1}{2}(\rho''^2 - 1)$, with $\rho' - \rho'' \equiv n \bmod. 2$, $|\rho' - \rho''| \leqq n$; when they are so, then, putting $h = \frac{1}{2}(n - \rho' + \rho'')$, we have the solution given by $f_h(p) = p^{1+(\rho'+\rho'')/2}$, $f_j = 0$ for $j \neq h$. In addition to this, for $\rho' = \rho'' = 0$, we have a solution $f_{n/2}(p) = p \log \frac{1}{p}$, $f_j = 0$ for $j \neq n/2$. According to the values of δ', δ'', n, this gives 0, 2 or 4 linearly independent solutions.

40. Ultimately, among harmonic functions, only those which are B-moderate will be of value to us. As to these, a first basic result is the following:

Proposition 9. Take $\xi \in K^\times$; consider on G^o the functions φ other than 0 which are B-moderate, harmonic of type (α, δ, ν) resp. $(\alpha, \delta', \delta'', n)$ and satisfy the condition

$$(25) \qquad \varphi\left(\begin{pmatrix}1 & y\\ 0 & 1\end{pmatrix}.g\right) = \psi(\xi y)\varphi(g)$$

for all $y \in K$ and $g \in G$. Then, up to a constant factor, there is at most one such function; there is one if $K = \mathbb{R}$; if $K = \mathbb{C}$, there is none unless δ', δ'' are of the form

$$\delta' = \frac{1}{2}(\rho'^2 - 1), \quad \delta'' = \frac{1}{2}(\rho''^2 - 1)$$

with $\rho' - \rho'' \equiv n \bmod. 2$, $|\rho' - \rho''| \leqq n$.

On G^o, as we have seen, a function of the given type is uniquely determined by the function $f(p, y)$ it induces on B_1; with the additional conditions we are imposing here, this must be of the form

$$f(p, y) = \psi(\xi y)f(p)$$

with $f(p) = O(p^N)$ for some N, for $p \longrightarrow +\infty$. Taking now first the case $K = \mathbb{R}$ and making use of (23), we see that f defines a solution if and only if it satisfies the equation

$$\frac{d^2 f}{dp^2} + \left(-4\pi^2\xi^2 - \frac{2\pi\nu\xi}{p} - \frac{\delta}{2p^2}\right)f = 0 .$$

After putting $z = 4\pi|\xi|p$, $f(p) = W(z)$, $\kappa = -\frac{\nu}{2}\operatorname{sgn}\xi$, $\rho = (1 + 2\delta)^{1/2}$, $\mu = \rho/2$, this becomes the well-known equation for the "confluent hypergeometric function" in standard form:

$$\frac{d^2 W}{dz^2} + \left(-\frac{1}{4} + \frac{\kappa}{z} + \frac{1/4 - \mu^2}{z^2}\right) W = 0 .$$

Of the two linearly independent solutions of this equation, it is well-known[1] that one, "Whittaker's function" $W_{\kappa,\mu}(z)$, tends to 0 exponentially for $z \longrightarrow +\infty$, while the other increases exponentially for $z \longrightarrow +\infty$. This proves our assertions for $K = \mathbb{R}$. We recall that $W_{\kappa,\mu}$ is so normalized that $W_{\kappa,\mu}(z) \sim e^{-z/2} z^{\kappa}$ for $z \longrightarrow +\infty$, and that $W_{\kappa,\mu} = W_{\kappa,-\mu}$.

Now take $K = \mathbb{C}$. Then f must be a vector-valued function $(f_o, \ldots, f_n)$; using (24) and the similar formula for D'', we get two differential equations for f, viz.

---

(1) Cf. e.g. W. Magnus, F. Oberhettinger and R. P. Soni, Formulas and Theorems for the Special Functions of Mathematical Physics (Springer 1966), Chapter VII.

$$(E') \quad \frac{d^2f}{dp^2} + \frac{1}{p}\frac{df}{dp}\cdot(A_n - 1_{n+1}) - 16\pi^2\xi\bar{\xi}f + f\cdot\left(\frac{8\pi i\xi}{p}A'_n + \frac{(A_n/2)^2 - A_n - 2\delta'.1_{n+1}}{p^2}\right) = 0$$

and the equation (E'') obtained by replacing ξ, $\bar{\xi}$, A_n, A'_n, δ' by $\bar{\xi}$, ξ, $-A_n$, A''_n, δ'', respectively, in (E'). As we have noted, A_n is the diagonal matrix with the coefficients $n-2j$; putting $A'_n = (a'_{hj})$, $A''_n = (a''_{hj})$ with $0 \leqq h, j \leqq n$, one finds at once that a'_{hj} is $-h$ if $j-h = -1$, and 0 otherwise, and a''_{hj} is $n-h$ if $j-h = 1$, and 0 otherwise. In particular, taking the n-th component in (E'), we get

$$\frac{d^2f_n}{dp^2} - \frac{n+1}{p}\frac{df_n}{dp} + \left(-16\pi^2\xi\bar{\xi} + \frac{(n/2)^2 + n - 2\delta'}{p^2}\right)f_n = 0 .$$

If we write $z = 4\pi(\xi\bar{\xi})^{1/2}p$, $K(z) = p^{-(n/2)-1}f_n(p)$, $\mu = (1 + 2\delta')^{1/2}$, this gives

$$\frac{d^2K}{dz^2} + \frac{1}{z}\frac{dK}{dz} - (1 + \frac{\mu^2}{z^2})K = 0 ,$$

which is the equation for "Hankel's function" in standard form (cf. W. Magnus, etc., loc. cit., Chapter III). Hankel's function $K_\mu(z)$ is the only solution which does not increase exponentially for $z \longrightarrow +\infty$; it is so normalized that $K_\mu(z) \sim e^{-z}(2z/\pi)^{-1/2}$ for $z \longrightarrow +\infty$, and $K_\mu = K_{-\mu}$. Thus, up to a constant factor, f_n is uniquely determined. Now take the j-th component of (E'') for $0 < j \leqq n$; this is of the form

$$\frac{d^2f_j}{dp^2} - \frac{n - 2j + 1}{p}\frac{df_j}{dp} + (\alpha_j + \frac{\beta_j}{p^2})f_j = -\frac{8\pi i\bar{\xi}}{p}(n-j+1)f_{j-1} ,$$

with certain constant coefficients α_j, β_j. By induction on $n-j$, this shows that f_{n-1}, f_{n-1}, ..., f_o are uniquely determined in terms of f_n.

This proves the unicity of the solution, if one exists. Furthermore, near $z = 0$, all solutions of the equation for K_μ are of the form $z^\mu \mathfrak{p}(z^2) + z^{-\mu}\mathfrak{q}(z^2)$ if μ is not in $\mathbf{Z}$, and $z^{-\mu}\mathfrak{p}(z^2) + z^\mu \log\frac{1}{z}\mathfrak{q}(z^2)$ if μ is an integer $\geqq 0$, $\mathfrak{p}$ and $\mathfrak{q}$ being power-series. From this and the equation for f_j obtained above, it follows at once that, for μ not in $\mathbf{Z}$, we can write f in the form $p^\alpha F(p) + p^\beta G(p)$, with $\alpha \not\equiv \beta$ mod. 2, $F = (F_o, \ldots, F_n)$ and $G = (G_o, \ldots, G_n)$ holomorphic near $p = 0$, F_j and G_j being even or odd functions of p according as j is even or odd; substituting this for f in (E'), (E''), we see that $p^\alpha F(p)$ and $p^\beta G(p)$ must both be solutions of (E'), (E''). Similarly, if μ is in $\mathbf{Z}$, we can write f as $p^\alpha[F(p) + \log\frac{1}{p}G(p)]$, with F, G as before, and find that $p^\alpha G(p)$ must be a solution of (E'), (E''), while $p^\alpha F(p)$ must clearly be one if $G = 0$. Thus, in all cases, if $f \neq 0$, there must be a solution of (E') and (E'') of the form $p^\alpha F(p)$, with F as above and not 0. Now let p^ν be the lowest power of p occurring in all the power-series for the functions F_j, so that $F(p) = cp^\nu + p^{\nu+1}G(p)$, with $c = (c_o, \ldots, c_n) \neq 0$ and G holomorphic; put $\lambda = \alpha + \nu$. Then one sees at once that cp^λ is a solution of the equations (E'_o), (E''_o) of §39, c). Now, if (e.g.) $c_h \neq 0$, we must have

$$(\lambda - 1 + \frac{n-2h}{2})^2 = 1 + 2\delta' \quad , \quad (\lambda - 1 - \frac{n-2h}{2})^2 = 1 + 2\delta'' \ .$$

Writing now ρ', ρ'' for $\lambda - 1 \pm (n-2h)/2$, we see that δ', δ'', n have the form asserted by our proposition.

Corollary. Let φ be as in proposition 9; for $g = p^{-1/2}\begin{pmatrix} p & 0 \\ 0 & 1 \end{pmatrix}$, put $\varphi(g) = f(p)$. Then $f(p) = O(e^{-Ap})$, with a suitable $A > 0$, for $p \longrightarrow +\infty$, and $f(p) = O(p^{-B})$, with a suitable B, for $p \longrightarrow 0$, and so are all its derivatives.

The first assertion follows at once from the behavior of Whittaker's function for $K = \mathbf{R}$, of Hankel's function for $K = \mathbf{C}$, and from that of their derivatives. For $K = \mathbf{C}$, we have shown that $f(p)$ is

$O(p^\lambda)$, or at worst $O(p^\lambda \log \frac{1}{p})$, λ being one of the finitely many exponents for which (E'_o), (E''_o) have a solution of the form cp^λ. For quite similar reasons, an analogous statement holds true for $K = \mathbf{R}$.

Remark 1. Pursuing the argument in the proof of our proposition a little further, one could verify the existence of a solution, also for $K = \mathbf{C}$; as this would require some computation, and as another proof will be given in §48, we omit it here.

Remark 2. Actually, for $K = \mathbf{C}$, our argument proves more than we have stated; it shows that there is no harmonic function of the given type (B-moderate or not), satisfying (25), with $\xi \neq 0$ or $\xi = 0$, unless δ', δ'', n are as stated there. One may ask whether there is <u>any</u> harmonic function of that type if they are otherwise; at any rate, our argument shows that such a function, if it exists, cannot have a Fourier transform (as a function of y) and cannot be periodic in y (in which case it would have a Fourier series).

41. If $K = \mathbf{C}$, we will say that a type $(\alpha, \delta', \delta'', n)$ is <u>admissible</u> if α, n satisfy the compatibility condition $\alpha(-1) = (-1)^n$ and if δ', δ'', n have the form stated in proposition 9. By an <u>admissible function</u> of such a type, we will then understand any B-moderate harmonic function of that type. Proposition 9 says then that, for each admissible type and each $\xi \neq 0$, there is (up to a constant factor) at most one admissible function, other than 0, satisfying (25); it will be seen in §48 that there does exist one.

For $K = \mathbf{R}$, things are less simple, because G is not connected; in order to define "admissible functions" (a definition without which we could not even state the "local functional equation" in the form in which it will be needed), we proceed as follows.

Put $W = U - i(X + Y)$; this is the left-invariant differential operator on G, corresponding to the matrix $\begin{pmatrix} 1 & -i \\ -i & -1 \end{pmatrix}$ of the complexified Lie algebra of G. We have $[X - Y, W] = -2iW$, which can also be written as

$$(X - Y).W = W.(X - Y - 2i) \ .$$

Since the functions φ on G, or on G^o, which satisfy the condition $\varphi(g.r(\theta)) = \varphi(g)e^{i\nu\theta}$ can be characterized by $(X - Y)\varphi = i\nu\varphi$, this shows that, if φ is such, $W\varphi$ and $\overline{W}\varphi$ have the similar property, with ν replaced by $\nu - 2$ and by $\nu + 2$, respectively.

For each integer ν, write H_ν for the space of harmonic functions of type (α, δ, ν). The above property of W, $\overline{W}$, together with the fact that D is central (and therefore commutes with W, $\overline{W}$) shows that W maps H_ν into $H_{\nu-2}$ and that $\overline{W}$ maps $H_{\nu-2}$ back into H_ν. An easy calculation shows that we have

(26) $$\overline{W}W = 2D + (X - Y)^2 - 2i(X - Y) \ ;$$

therefore $\overline{W}W$, applied to functions in H_ν, multiplies them with the scalar factor $2\delta - \nu^2 + 2\nu$; in particular, W and $\overline{W}$ are isomorphisms of H_ν and $H_{\nu-2}$ onto one another if (and only if) that factor is not 0.

As W and $\overline{W}$ are left-invariant, they transform any function satisfying (25) into another such function; applying this to the results proved in proposition 9, one gets recurrence relations for the Whittaker functions, which of course are well-known.

Now we put again $\rho = (1 + 2\delta)^{1/2}$, where for definiteness we choose the square root so that $Re(\rho) > 0$, or $Im(\rho) \geqq 0$ if $Re(\rho) = 0$. An easy induction shows that, for each $i > 0$, the operator $\overline{W}^i W^i$, applied to the functions in H_ν, multiplies them with the scalar factor

(27) $$\gamma_{i,\nu} = \prod_{1 \leqq j \leqq i} [\rho^2 - (\nu - 2j+1)^2] \ .$$

In particular, for any $n \geqq 0$, the operator $\overline{W}^n W^n$, applied to the functions in H_n, multiplies them with the factor $\gamma_{n,n} = C_n^2$, with

$C_o = 1$ and

$$C_n = (\rho + n - 1)(\rho - n - 3) \dots (\rho - n + 1)$$

for $n > 0$; $W^n \overline{W}^n$, applied to functions in H_{-n}, multiplies them with the same factor. We have $C_n = 0$ if $\rho \in \mathbf{Z}$, $n > \rho$, $n \equiv \rho + 1 \bmod. 2$; otherwise C_n is not 0, and then the operators $C_n^{-1} W^n$, $C_n^{-1} \overline{W}^n$ are isomorphisms of H_n onto H_{-n}, and of H_{-n} onto H_n, respectively, and are the inverses of each other. One may note that, if ρ were replaced by $-\rho$, C_n would be changed into $(-1)^n C_n$.

For $C_n \neq 0$, we will say that a function φ in H_n, and a function ψ in H_{-n}, are <u>conjugate</u> if $\psi = C_n^{-1} W^n \varphi$ (and consequently $\varphi = C_n^{-1} \overline{W}^n \psi$); when that is so, we will write $\psi = \tilde{\varphi}$, $\varphi = \tilde{\psi}$. Observe that, if s is the matrix $\begin{pmatrix} -1 & 0 \\ 0 & 1 \end{pmatrix}$ as before, and φ is in H_n, the function φ_1 given by $\varphi_1(g) = \varphi(gs)$ is in H_{-n}. Moreover, W and $\overline{W}$ are interchanged by the inner automorphisms $g \longrightarrow s^{-1} gs$, and therefore also by $s \longrightarrow gs$ (since they are left-invariant). Consequently, if φ, φ_1 are as above, the conjugate of φ_1 is given by $\tilde{\varphi}_1(g) = \tilde{\varphi}(gs)$. Assume now that φ has been given only on the connected component G^o and belongs to H_n there; take $d = 0$ or 1, put $\psi = (-1)^d \tilde{\varphi}$ on G^o, and define φ on the other component $G^o s$ by taking $\varphi(gs) = \psi(g)$ for $g \in G^o$. Then φ still belongs to H_n on $G^o s$, and, if we take also $\psi = (-1)^d \tilde{\varphi}$ on $G^o s$, we have $\psi(g) = \varphi(gs)$ and $\varphi(s) = \psi(gs)$ for $g \in G$.

We are now ready to define admissible types and admissible functions for $K = \mathbf{R}$. We distinguish several cases:

a) For $n > 0$, $\alpha(-1) = (-1)^n$, $C_n \neq 0$, $d = 0$ or 1, we will say that (α, δ, d, n) is an <u>admissible type</u>. By an <u>admissible function</u> $\Phi = (\varphi_1, \varphi_2)$ of that type, we will understand a B-moderate harmonic function of type (α, δ, M_n) such that $\varphi_2 = (-1)^d \tilde{\varphi}_1$. Here M_n is again the representation of $\mathfrak{K}$ of degree 2 defined in §35, a). From what has been said above, it follows that φ_1, φ_2 are well determined on the whole of G as soon as one of them is given on G^o.

b) For $n = 0$, $\alpha(-1) = 1$, $d = 0$ or 1, we still say that $(\alpha, \delta, d, 0)$ is an admissible type. By an admissible function Φ of that type, we understand a B-moderate $\mathbf{C}$-valued harmonic function of type (α, δ, M_o) or of type (α, δ, M'_o) according as d is 0 or 1; here M_o, M'_o are again as in §35, a). In other words, Φ is to be right-invariant under $\mathfrak{K}_1$ and to satisfy $\Phi(gs) = (-1)^d \Phi(g)$, so that it is again determined by its values on G^o.

In the two cases a), b) (which are those for which $C_n \neq 0$), we say that (α, δ, d, n) is a principal type, while the types to be defined in c) (those for which $C_n = 0$) will be called discrete. This terminology is suggested by the theory of representations, although it does not quite agree with it (the case $\rho \equiv n + 1 \bmod. 2$, $\rho > n$ belongs neither to the "principal series" nor to the "discrete series" of that theory).

c) For $C_n = 0$, we must still have $\alpha(-1) = (-1)^n$, of course, and, with ρ defined as above, we have $n = \rho + 2h + 1$, where h is an integer $\geqq 0$; when that is so, we call (α, δ, n) an admissible discrete type. By an admissible function $\Phi = (\varphi_1, \varphi_2)$ of that type, we will understand a B-moderate harmonic function of type (α, δ, M_n) such that $W^{h+1}\varphi_1 = 0$. As our conditions imply $\varphi_2(g) = \varphi_1(gs)$, we have $\overline{W}^{h+1}\varphi_2 = 0$.

Remark 1. In all three cases, the admissible function Φ is uniquely determined by its first component φ_1 (taking this to be the same as Φ for $n = 0$); this is B-moderate and harmonic of type (α, δ, n), and we have $W^n\varphi_1 = (-1)^d C_n \varphi_2$, also in case c). Indeed, considering discrete types as limits of principal types, it might seem more natural to define admissible functions, in case c), by $W^n\varphi_1 = 0$

rather than $W^{h+1}\varphi_1 = 0$. As may be seen from the proof of proposition 10, this would merely enlarge the space of admissible functions by a finite-dimensional space which would then be discarded automatically once one considers only automorphic functions. From the point of view of representation theory, this corresponds to the fact that a certain representation decomposes into a finite-dimensional one and an irreducible one of infinite dimension.

Remark 2. For $K = \mathbb{C}$, the fact that we have chosen the representatives M_n for the classes of irreducible representations of $\mathfrak{k}_1$ plays no essential role in our definition of admissible functions; if M is any such representation, we can say that a function Φ belonging to M is admissible if it is B-moderate and harmonic of some type $(\alpha, \delta', \delta'')$. As M must be equivalent to one of the representations M_n, we can write $M = A^{-1}.M_n.A$, and it is clear that Φ is admissible in this new sense if and only if ΦA^{-1} is so in the former sense. For $K = \mathbb{C}$, our definition of admissible functions has also the following invariance property: let F be a matrix-valued function on G with compact support, such that $F(\mathfrak{k} g \mathfrak{k}') = M_n(\mathfrak{k})^{-1}F(g)M_{n'}(\mathfrak{k}')$; then it is obvious that, if Φ is admissible of type $(\alpha, \delta', \delta'', n)$, the convolution $\Phi * F$ is admissible of type $(\alpha, \delta', \delta'', n')$. Presumably a similar property holds for $K = \mathbf{R}$, but our definition does not make it obvious; moreover, for $n > 0$, our definition of admissible functions is tied up with our special choice of representatives M_n for the classes of representations of $\mathfrak{k}$, and would have to be modified if another choice was made.

42. For $K = \mathbf{R}$, proposition 9 makes it clear that there is, within each admissible principal type, only one admissible function (up to a constant factor) satisfying (25). The same conclusion will be seen to hold for discrete types after we prove the following:

Proposition 10. *Let* (α, δ, n) *be a discrete type, so that*

$n = \rho + 2h + 1$ with $h \in \mathbf{Z}$, $h \geqq 0$; let φ be harmonic of type (α, δ, n) on G^o. Then, if φ is B-moderate and satisfies (25), we have $W^{h+1}\varphi = 0$ whenever $\xi < 0$, while if $\xi > 0$, $W^n\varphi \neq 0$ unless $\varphi = 0$.

To begin with, let φ be any function on G^o, satisfying $\varphi(g.r(\theta)) = \varphi(g)e^{i\nu\theta}$; call f, i.e. $(p, y) \longrightarrow f(p, y)$, the function induced by φ on B_1; put $\tau = y + ip$. An easy calculation shows that the functions induced on B_1 by $W\varphi$ and by $\overline{W}\varphi$ are $-4ip\frac{\partial f}{\partial \bar{\tau}} - \nu f$ and $4ip\frac{\partial f}{\partial \tau} + \nu f$, respectively. From this one concludes, by induction on r, that the function induced on B_1 by $W^r\varphi$ is

$$f_r(p, y) = (-4i)^r p^{1+(\nu/2)} \cdot \frac{\partial^r}{\partial \bar{\tau}^r}[p^{r-1-(\nu/2)}f] .$$

Now consider the discrete type (α, δ, n); as ρ was defined as $(1 + 2\delta)^{1/2}$, and as $\rho = n - 2h - 1$, we have $2\delta = m^2 - m$ with $m = n - 2h$. We have seen that W, $\overline{W}$ define isomorphisms of H_ν, $H_{\nu-2}$ onto one another unless $2\delta = \nu^2 - 2\nu$, i.e. here unless $\nu = m$ or $\nu = 2 - m$. Take any function φ in H_n; then $\varphi' = W^h\varphi$ is in H_m, and, since the constant $\gamma_{h,n}$ defined by (27) is not 0, we have $\varphi = \gamma_{h,n}^{-1}\overline{W}^h\varphi'$. Moreover, (26) shows that we have $\overline{W}W\varphi' = 0$. Put $\varphi'' = W\varphi'$; this is in H_{m-2}, and $\overline{W}\varphi'' = 0$; if f'' is the function induced by φ'' on B_1, the formulas given above show that this can be written as $\frac{\partial}{\partial \tau}(p^{(m/2)-1}f'') = 0$, so that we can write $f''(p, y) = p^{1-(m/2)}F(\bar{\tau})$, where F is a holomorphic function of $\bar{\tau} = y - ip$ in the lower half-plane. Now we have $W^n\varphi = W^{m+h-1}\varphi''$; as $W^{m-1}\varphi''$ is in H_{-m}, and as W maps H_ν isomorphically onto $H_{\nu-2}$ for all $\nu \leqq -m$, $W^n\varphi$ is 0 if and only if $W^{m-1}\varphi'' = 0$; in view of the formula given above for $f_r(p, y)$ (where we have to substitute $m-1$ for r, $m-2$ for ν, $p^{1-(m/2)}F(\bar{\tau})$ for f), this amounts to

$\frac{d^{m-1}F}{d\bar{\tau}^{m-1}} = 0$; consequently, $W^n\varphi$ is 0 if and only if F is a polynomial in $\bar{\tau}$, of degree $< m-1$. Now assume that φ satisfies (25) for some $\xi \neq 0$; as W is left-invariant, the same must hold for φ'', which is clearly impossible if F is a polynomial $\neq 0$ in $\bar{\tau}$. Therefore, if φ is as in our proposition, $W^n\varphi$ is 0 if and only if $W\varphi' = 0$, i.e. $W^{h+1}\varphi = 0$. But, if f' is the function induced by φ' on B_1, this can be written as $\frac{\partial}{\partial \tau}(p^{-m/2}f') = 0$; this is so if and only if $f' = p^{m/2}F'(\tau)$, with F' holomorphic in the upper half-plane. If this satisfies (25), we must have $F'(\tau + b) = e^{-2\pi i\xi b}F'(\tau)$ for all $b \in \mathbf{R}$; for F' holomorphic, this gives $F'(\tau) = C\,e^{-2\pi i\xi\tau}$, with a constant C. Clearly this is B-moderate if and only if $\xi < 0$; as φ, up to a constant factor, is the same as $\overline{W}^h\varphi'$, it is easily seen that the same is true of φ. Therefore, if $\xi > 0$, $W^n\varphi = 0$ implies $\varphi = 0$, as asserted by our proposition. On the other hand, if $\xi < 0$, take F', f', φ', φ as we have just said; then φ has all the properties described in proposition 9 and is not 0, and it satisfies $W^{h+1}\varphi = 0$; in view of the unicity of such a function, this completes the proof.

Corollary. For each admissible type for $K = \mathbf{R}$, and each $\xi \in K^\times$, there is, up to a constant factor, one and only one admissible function Φ on G, other than 0, satisfying (25).

For a principal type, this follows at once from proposition 9 and the definitions, together with the fact that (since W is left-invariant), $\tilde{\varphi}$ satisfies (25) if φ does. For a discrete type, it follows from proposition 10; in fact, taking such a type and putting $\Phi = (\varphi_1, \varphi_2)$, we see that φ_1 is 0 on one of the components G^o, $G^o s$ of G, and is given by proposition 9 on the other, while φ_2 is given by $\varphi_2(gs) = \varphi_1(g)$ for all $g \in G$. More precisely, proposition 10 shows that φ_1 is 0 on G^o if $\xi > 0$, and on $G^o s$ if $\xi < 0$; moreover, the formula for $\overline{W}^h\varphi'$, similar to the one given for $W^r\varphi$ in the proof of proposition 10, shows that, for $\xi < 0$, the function f_1 induced by φ_1 on B_1 is given by

$$f_1(p, y) = p^{1-(m/2)} \frac{d^h}{dp^h}(p^{n-h-1}e^{-2\pi|\xi|p}) ,$$

while, for $\xi > 0$, the function induced by $g \longrightarrow \varphi_1(sg)$ is given by that same formula. This determines Φ completely; needless to say, it agrees with known facts about Whittaker's function.

43. In order to simplify the statements and proofs in the remainder of this Chapter, it is convenient to make two observations.

Firstly, let Φ be any function satisfying

$$\Phi(\begin{pmatrix} 1 & y \\ 0 & 1 \end{pmatrix} . g) = \psi(y)\Phi(g)$$

for all $y \in K$ and $g \in G$; then, for any $\xi \in K^\times$, the function $g \longrightarrow \Phi(\begin{pmatrix} \xi & 0 \\ 0 & 1 \end{pmatrix} . g)$ satisfies (25); if Φ is admissible (of any type), so is the latter function. Thus, in discussing admissible functions satisfying (25), it is enough to consider the case $\xi = 1$.

Secondly, take an admissible type $(\alpha, \delta', \delta'', n)$ in the case $K = \mathbf{C}$; let Φ be any admissible function of that type, and let λ be any quasicharacter of $K^\times$. Then $g \longrightarrow \Phi(g)\lambda(\det g)$ is admissible of type $(\alpha\lambda^2, \delta', \delta'', n)$. Let α be given by $\alpha(x) = x^{\zeta'}\bar{x}^{\zeta''}$; as the given type is admissible, we have $\zeta' - \zeta'' \equiv n \bmod. 2$, and we may write $\delta' = \frac{1}{2}(\rho'^2 - 1)$, $\delta'' = \frac{1}{2}(\rho''^2 - 1)$, with $\rho' - \rho'' \equiv n \bmod. 2$, $|\rho' - \rho''| \leqq n$; here, after replacing ρ', ρ'' by $-\rho'$, $-\rho''$, we may assume that $\mathrm{Re}(\rho' + \rho'') \geqq 0$. Put now $\sigma' = \frac{1}{2}(\rho' - \zeta' - 1)$, $\sigma'' = \frac{1}{2}(\rho'' - \zeta'' - 1)$; as $\sigma' \equiv \sigma'' \bmod. 1$, we can define a quasicharacter λ of $K^\times$ by $\lambda(x) = x^{\sigma'}\bar{x}^{\sigma''}$; then we have $(\alpha\lambda^2)(x) = x^{\rho'-1}\bar{x}^{\rho''-1}$. Thus, by modifying Φ in the manner indicated, we can always change it into one of a type $(\alpha, \delta', \delta'', n)$ for which we have:

$$(28) \quad \begin{cases} \delta' = \frac{1}{2}(\rho'^2 - 1), \ \delta'' = \frac{1}{2}(\rho''^2 - 1), \ \alpha(x) = x^{\rho'-1}\bar{x}^{\rho''-1} , \\ \rho' - \rho'' \equiv n \bmod. 2, \ |\rho' - \rho''| \leqq n, \ \mathrm{Re}(\rho' + \rho'') \geqq 0 . \end{cases}$$

A type satisfying these conditions will be called reduced, except when ρ', ρ'' are integers > 0, and $\rho' + \rho'' \leqq n$; in the latter case, we carry the "reduction" one step further, so as to replace α by $\alpha\lambda^2$ with $\lambda(x) = x^{-\rho'}$ if $\rho' \leqq \rho''$, and $\lambda(x) = \bar{x}^{-\rho''}$ if $\rho'' < \rho'$. Thus a type will be called reduced if it satisfies (28), with the additional condition, if ρ', ρ'' are integers, that either $\rho'\rho'' \leqq 0$ or $\rho' + \rho'' > n$. Since a reduced type $(\alpha, \delta', \delta'', n)$ is uniquely determined by α and n, we will denote it by $[\alpha, n]$.

For $K = \mathbf{R}$, the above procedure must be slightly modified. As above, take a quasicharacter λ of $K^\times$. If Φ is admissible of type $(\alpha, \delta, d, 0)$, it is $\mathbf{C}$-valued, and we again consider the function $g \longrightarrow \Phi(g)\lambda(\det g)$; this is admissible of type $(\alpha\lambda^2, \delta, d', 0)$, with $(-1)^{d'} = (-1)^d\lambda(-1)$. For a discrete type (α, δ, n) or a principal type (α, δ, d, n) with $n > 0$, we can write $\Phi = (\varphi_1, \varphi_2)$; in that case, we consider the function

$$g \longrightarrow \lambda(\det g).(\varphi_1(g), \lambda(-1)\varphi_2(g)) \ ;$$

it is easily verified that this is admissible of type $(\alpha\lambda^2, \delta, n)$ resp. $(\alpha\lambda^2, \delta, d', n)$, again with $(-1)^{d'} = (-1)^d\lambda(-1)$. Taking $\rho = (1 + 2\delta)^{1/2}$ with $\mathrm{Re}(\rho) \geqq 0$, and $\sigma = \frac{1}{2}(\rho - \zeta - 1)$, we take now $\lambda(x) = |x|^\sigma$ in the case of a discrete type, and $\lambda(x) = (\mathrm{sgn}\, x)^{n-d}|x|^\sigma$ for a principal type. Thus Φ is changed into an admissible function whose type (α, δ, n) resp. (α, δ, d, n) is such that

$$(29)\quad \begin{cases} \delta = \frac{1}{2}(\rho^2 - 1),\ \alpha(x) = (\mathrm{sgn}\, x)^n |x|^{\rho-1},\ \mathrm{Re}(\rho) \geqq 0 \ , \\ \text{(a) } n \equiv \rho + 1 \text{ mod. } 2,\ n > \rho; \ \underline{\text{or}} \text{ (b) } n \equiv d \text{ mod. } 2 \ ; \end{cases}$$

(a) and (b) correspond to the cases of a discrete type and of a principal type, respectively. Such a type will be called reduced; as it depends only upon α and n, we will denote it by $[\alpha, n]$.

44. To simplify the language, we shall understand by the standard function of a given type the admissible function Φ of that type, other than 0, which satisfies

$$\Phi(\begin{pmatrix} 1 & y \\ 0 & 1 \end{pmatrix} . g) = \psi(y)\Phi(g) \tag{30}$$

for all $y \in K$ and all $g \in G$. From proposition 9, and the corollary of proposition 10, we know that this function (if it exists) is uniquely defined up to a constant factor. For such functions, we will give an integral representation which will not only fill up the gap in the existence proof for $K = \mathbb{C}$, but, more significantly, will lead at once to the "local functional equation" of theorem 4 below.

For the reasons explained in §43, it will be enough to deal with reduced types. Let therefore $[\alpha, n]$ be such a type, and take notations again as in (28) resp. (29). The function φ of §39, a), given by $\varphi(g) = f(\begin{pmatrix} a \\ c \end{pmatrix}) = \alpha(c)\psi\left(\frac{a}{c}\right)$ for $g = \begin{pmatrix} a & b \\ c & d \end{pmatrix}$, is then harmonic of type (α, δ) resp. $(\alpha, \delta', \delta'')$ wherever it is regular, i.e. for $c \neq 0$, and it satisfies (30); the same is therefore true of all its right-translates, so that we may hope to build up admissible functions by convoluting φ with suitable mass-distributions. This will be done now.

Consider first any right-translate $g \longrightarrow \varphi(gg')$ of φ; for $g' = \begin{pmatrix} a' & b' \\ c' & d' \end{pmatrix}$, this can be written as $g \longrightarrow f(g.\begin{pmatrix} a' \\ c' \end{pmatrix})$, where $g.\begin{pmatrix} a' \\ c' \end{pmatrix}$ is understood in the sense of matrix multiplication. Consequently, if we take for S (to begin with) a continuous function with compact support in $K^2 - \{0\}$, the function

$$\Psi_S(g) = \int f(g.\begin{pmatrix} x \\ y \end{pmatrix})S(\begin{pmatrix} x \\ y \end{pmatrix})dxdy$$

is a right-convolute of φ on G and may be expected to have the properties described above.

Proceeding formally at first, take for S any "Schwartz function" in K^2, i.e. a function of class C^∞, which, together with all its derivatives, tends to 0 at infinity faster than any power $(x\bar{x} + y\bar{y})^{-N}$; a typical case (one which would actually suffice for our purposes) is given by

$$S\left(\binom{x}{y}\right) = P(x, \bar{x}, y, \bar{y})e^{-x\bar{x}-y\bar{y}} ,$$

where P is a polynomial; of course, if $K = \mathbf{R}$, it is understood that $x = \bar{x}$ and $y = \bar{y}$. Put

$$S_g(x, y) = S\left(g^{-1}\binom{x}{y}\right) .$$

Formally, the above integral can also be written

$$\begin{aligned}\Psi_S(g) &= |\det g|_K^{-1} \int f\left(\binom{x}{y}\right) S_g(x, y)dxdy \\ &= |\det g|_K^{-1} \int \alpha(y)\psi\left(\frac{x}{y}\right) S_g(x, y)dxdy .\end{aligned}$$

As $|\alpha(y)|$ is of the form $|y|_K^\sigma$, this will be absolutely convergent if $\sigma > -1$, i.e. (with the notations of §43) if $\mathrm{Re}(\rho) > 0$ resp. $\mathrm{Re}(\rho' + \rho'') > 0$; then we can also write:

$$\Psi_S(g) = |\det g|_K^{-1} \int_{K^\times} \left[\int_K S_g(x, y)\psi\left(\frac{x}{y}\right)dx\right] \alpha(y)dy . \tag{31}$$

As S_g is a Schwartz function, it has a "partial Fourier transform":

$$S'_g(u, y) = \int_K S_g(x, y)\psi(xu)dx$$

which is also a Schwartz function; more precisely, since $g \longrightarrow S_g$ is clearly a differentiable mapping of G into the space of Schwartz functions on K^2, so is $g \longrightarrow S'_g$. Consequently the integrand in (31), which can be written as $S'_g(1/y, y)\alpha(y)$, is $O(|y|_K^{-N})$ for $|y|_K \longrightarrow +\infty$, and $O(|y|_K^N)$ for $|y|_K \longrightarrow 0$, for all N, and (31) is always absolutely convergent;

more precisely, the formula

$$S' \longrightarrow A(S') = \int_{K^\times} S'(1/y,\ y)\alpha(y)dy$$

defines a temperate distribution A in K^2, and (31) may be written as

(32) $$\Psi_S(g) = |\det g|_K^{-1} A(S'_g) \ .$$

45. We can now prove the following:

Proposition 11. For every Schwartz function S, the function Ψ_S defined by (31) satisfies (30) and is B-moderate and harmonic of type (α, δ) resp. $(\alpha, \delta', \delta'')$, where δ is given by (29) (resp. δ', δ'' by (28)); it is admissible of type $[\alpha, n]$ if S satisfies $S(\mathfrak{p}^{-1}\begin{pmatrix} x \\ y \end{pmatrix}) = S(\begin{pmatrix} x \\ y \end{pmatrix})M_n(\mathfrak{p})$ for all $\mathfrak{p} \in \mathfrak{K}$.

Replace g by $\begin{pmatrix} p & 0 \\ 0 & 1 \end{pmatrix}.g$ in (31), with $p > 0$, and then x by px; we get:

$$\Psi_S(\begin{pmatrix} p & 0 \\ 0 & 1 \end{pmatrix}.g) = |\det g|_K^{-1} \int_{K^\times} S'_g(p/y,\ y)\alpha(y)dy \ .$$

For g in a compact subset of G, and for any N, there is a constant γ such that $|S'_g(u,\ y)| \leq \gamma(u\bar{u} + y\bar{y} + 1)^{-N}$; therefore the integral is $O(p^{-N})$ for all N, for $p \longrightarrow +\infty$; in particular, Ψ_S is B-moderate. Replacing g by $\begin{pmatrix} 1 & u \\ 0 & 1 \end{pmatrix}.g$ in (31), and then x by $x + uy$, we see that Ψ_S satisfies (30). Replacing g by $g\mathfrak{z}$ with $\mathfrak{z} = z.1_2$ and then x, y by zx, zy, we see that Ψ_S belongs to the character α of $\mathfrak{Z}$; in particular, it is an eigenfunction of Z (resp. Z', Z'') for the eigenvalue $\rho - 1$ given by (29) (resp. for the eigenvalues $\rho' - 1$, $\rho'' - 1$ given by (28)).

Furthermore, it follows from basic results in the theory of temperate distributions that the formulas for Ψ_S may be differentiated formally with respect to g, so that Ψ_S is indefinitely differentiable on G. In order to make this explicit, we observe, e.g. for $K = \mathbb{R}$, that the operator X, applied to $g \longrightarrow S(g^{-1}\binom{x}{y})$ for fixed x, y, gives the function $g \longrightarrow S_X(g^{-1}\binom{x}{y})$ with $S_X = -y\frac{\partial S}{\partial x}$; similarly, applying Y, U, Z to the same function, we get functions $g \longrightarrow S_Y(g^{-1}\binom{x}{y})$, etc. with

(33) $$S_X = -y\frac{\partial S}{\partial x},\ S_Y = -x\frac{\partial S}{\partial y},\ S_U = -x\frac{\partial S}{\partial x} + y\frac{\partial S}{\partial y},\ S_Z = -x\frac{\partial S}{\partial x} - y\frac{\partial S}{\partial y}.$$

For $K = \mathbf{C}$, we have similar formulas, with (formally) the same right-hand sides as in (33) for $S_{X'}$, etc., and with $\bar{x}, \bar{y}$ substituted for x, y in these right-hand sides for $S_{X''}$, etc.

From this, one concludes in particular that $g \longrightarrow S(g^{-1}\binom{x}{y})$ is annulled by the operator $D - \frac{1}{2}Z^2 + Z$ (resp. by $D' - \frac{1}{2}Z'^2 + Z'$, $D'' - \frac{1}{2}Z''^2 + Z''$); consequently the same is true of $g \longrightarrow S'_g$ and of $g \longrightarrow A(S'_g)$. Then (32) shows that Ψ_S is annulled by $D - \frac{1}{2}Z^2 - Z$ (resp. by $D' - \frac{1}{2}Z'^2 - Z'$, $D'' - \frac{1}{2}Z''^2 - Z''$); as it is an eigenfunction of Z (resp. Z', Z'') for the eigenvalues mentioned above, it is therefore an eigenfunction of D (resp. D', D'') as required by our proposition.

The last assertion in our proposition is obvious for $K = \mathbf{C}$, and for $K = \mathbf{R}$ and $n = 0$. Take now the case $K = \mathbf{R}$, $n > 0$, and put $w = x + iy$. It is easily seen, then, that S has the required behavior under $\mathfrak{K}$ if and only if it is of the form

$$S(\binom{x}{y}) = E(w\bar{w}).(w^n, (-\bar{w})^n) ,$$

where E is a function on $\mathbf{R}_+$ such that $E(t^2)$ is a Schwartz function on $\mathbf{R}$. Call S_1, S_2 the two components of S, and put $t = w\bar{w}$, $w = t^{1/2}e^{i\varphi}$; then S_1 can be written as $F(t)e^{in\varphi}$, with $F(t) = E(t)t^{n/2}$.

With $W = U - i(X + Y)$ as in §41, (33) gives:

$$S_W = -2\bar{w}\frac{\partial S}{\partial w} = -(2t\frac{\partial S}{\partial t} - i\frac{\partial S}{\partial \varphi})e^{-2i\varphi},\ S_Z = -w\frac{\partial S}{\partial w} - \bar{w}\frac{\partial S}{\partial \bar{w}} = -2t\frac{\partial S}{\partial t}.$$

Writing, for convenience, $W(S)$, $Z(S)$ instead of S_W, S_Z, we get, by induction on h:

$$W^h(S_1) = (Z - n)(Z - n + 2) \ldots (Z - n + 2h - 2)S_1 . e^{-2ih\varphi} .$$

Take $h = n$; put

$$P(Z) = (Z - n)(Z - n + 2) \ldots (Z + n - 2) ;$$

we get:

$$W^n(S_1) = P(Z)S_1 . e^{-2in\varphi} = (-1)^n P(Z)S_2 .$$

As above, we conclude that a similar relation holds for the two components of $s \longrightarrow S'_g$ and for those of $s \longrightarrow A(S'_g)$; if then we write $\Psi_S = (\Psi_1, \Psi_2)$ and again make use of the fact that these are eigenfunctions of Z for the eigenvalue $\rho - 1$, we get now:

$$(-1)^n W^n \Psi_1 = P(Z + 2)\Psi_2 = P(\rho + 1)\Psi_2 = C_n \Psi_2 ,$$

with C_n as in §41. In view of (29), this completes the proof that Ψ_S is admissible of type $[\alpha, n]$, if that type is principal; if it is discrete, the same follows from the above result, in conjunction with proposition 10.

Corollary. Whenever S satisfies the condition in the latter part of proposition 11, Ψ_S is standard of type $[\alpha, n]$ provided it is not 0.

46. We will need "Tate's lemma" for K (cf. Tate's Thesis = Chapter XV of Algebraic Number Theory ("The Brighton Conference"), edd. J. W. S. Cassels and A. Fröhlich, Ac. Press 1967; cf. also the comments in A. Weil, Fonction zêta et distributions, Séminaire Bourbaki n° 312, Juin 1966). In order to formulate it, we define a topology and a complex structure on the group Ω_K of quasicharacters of $K^\times$, just as we have done in §9 in the global case, viz., by taking as connected component of 1 in Ω_K the group consisting of the quasicharacters

$\omega_s(x) = |x|_K^s$ with $s \in \mathbb{C}$, the complex structure being given by s; here $|x|_K$ is the "ordinary" absolute value if $K = \mathbf{R}$, and is $x\bar{x}$ if $K = \mathbb{C}$. For $K = \mathbf{R}$, Ω_K has two components, corresponding to $m = 0$ and to $m = 1$ in (21); for $K = \mathbb{C}$, (22) shows that there is one connected component, given by $\zeta' - \zeta'' = m$, for each $m \in \mathbf{Z}$. If ω is a quasicharacter, we may write $\text{abs}\,\omega(x) = \omega_\sigma(x) = |x|_K^\sigma$ with $\sigma \in \mathbf{R}$, and then we write $\sigma = \sigma(\omega)$. On K, we write dx for the additive Haar measure, normalized so that it is self-dual for $\psi(xy)$; this means that the Fourier transform on K is given by

$$\Phi^*(u) = \int_K \Phi(x)\psi(xu)dx\ , \quad \Phi(x) = \int_K \Phi^*(u)\psi(-xu)du\ .$$

We normalize the Haar measure $d^\times x$ on $K^\times$ by putting $d^\times x = |x|_K^{-1}dx$. On Ω_K, we introduce the following functions:

a) For $K = \mathbf{R}$, and α given by (21), we put

$$\mathcal{G}(\alpha) = G_1(\zeta + m) \text{ with } G_1(s) = \pi^{-s/2}\Gamma(s/2)\ .$$

For $K = \mathbb{C}$, and α given by (22), we write $\zeta' \geqq \zeta''$ or $\zeta' \leqq \zeta''$ according as $\zeta' - \zeta''$ is $\geqq 0$ or $\leqq 0$, and then we put

$$\mathcal{G}(\alpha) = G_2(\sup(\zeta', \zeta'')) \text{ with } G_2(s) = (2\pi)^{1-s}\Gamma(s)\ .$$

When necessary, we will write $\mathcal{G}_K$ instead of $\mathcal{G}$ for this function. Clearly its reciprocal $\mathcal{G}^{-1}$ is an entire (i.e. everywhere holomorphic) function on Ω_K, whose zeros are the quasicharacters $\alpha(x) = x^{-f}$ with $f \in \mathbf{Z}$, $f \geqq 0$, if $K = \mathbf{R}$, and $\alpha(x) = x^{-f'}\bar{x}^{-f''}$, with $f' \in \mathbf{Z}$, $f'' \in \mathbf{Z}$, $f' \geqq 0$, $f'' \geqq 0$, if $K = \mathbb{C}$.

b) With the same notations, we put $\kappa(\alpha) = i^{-m}$ if $K = \mathbf{R}$, and $\kappa(\alpha) = i^{-|\zeta' - \zeta''|}$ if $K = \mathbb{C}$; when necessary, we write κ_K for κ. This is locally constant (i.e. constant on each connected component of Ω_K);

we have $\kappa(1) = 1$, $\kappa(\alpha^{-1}) = \kappa(\alpha)$, and $\kappa(\alpha)^2 = \alpha(-1)$ for all α.

Tate's lemma is now as follows:

Lemma 6. _There is an everywhere holomorphic mapping_ $\alpha \longrightarrow \Delta_\alpha$ _of_ Ω_K _into the space of temperate distributions on_ K, _such that:_

(i) _For each_ α _and each_ $a \in K^\times$, _the transform of_ Δ_α _under the automorphism_ $x \longrightarrow ax$ _of_ K _is_ $\alpha(a)^{-1}\Delta_\alpha$; _conversely, every temperate distribution with that property is of the form_ $c\Delta_\alpha$ _with a constant_ c;

(ii) _If_ Φ _is a Schwartz function on_ K, _we have_

$$\Delta_\alpha(\Phi) = \mathcal{G}(\alpha)^{-1} \int_{K^\times} \Phi(x)\alpha(x)d^\times x$$

whenever the integral is absolutely convergent;

(iii) _If_ Φ _is as in (ii), and_ Φ^* _is its Fourier transform, we have_

$$\Delta_\alpha(\Phi) = \kappa(\alpha)^{-1}\Delta_{\omega_1\alpha^{-1}}(\Phi^*) \ .$$

For the integral in (ii) to converge for all Φ, $\sigma(\alpha)$ must be > 0; but the validity of (ii) is not restricted to that case. For instance, if the formal power-series for Φ at 0 has no terms of degree $< N$, the integral converges provided $\sigma(\alpha)$ is $> -N$ (for $K = \mathbf{R}$) or $> -N/2$ (for $K = \mathbf{C}$), and then (ii) is valid.

Some special cases need to be mentioned. For $\alpha = 1$, we have $\Delta_1(\Phi) = \Phi(0)$; more generally, Δ_α has the support $\{0\}$ if and only if α is a zero of $\mathcal{G}^{-1}$; more precisely, we have

$$\Delta_\alpha(\Phi) = \gamma_f\left(\frac{d^f\Phi}{dx^f}\right)_0 \quad \text{for } K = \mathbf{R},\ \alpha(x) = x^{-f},$$

$$\gamma_f = (-\pi)^{-e}\frac{e!}{f!},\quad e = [f/2]\ ;$$

$$\Delta_\alpha(\Phi) = \gamma_{f',f''}\left(\frac{\partial^{f'+f''}\Phi}{\partial x^{f'}\,\partial \bar{x}^{f''}}\right)_0 \quad \text{for } K = \mathbf{C},\ \alpha(x) = x^{-f'}\bar{x}^{-f''},$$

$$\gamma_{f',f''} = (-2\pi)^{-e}\frac{e!}{f'!\,f''!}$$

$$e = \inf(f',\ f'')\ .$$

For $K = \mathbf{R}$, $\Phi(x) = x^a e^{-\pi x^2}$, $\Delta_\alpha(\Phi)$ is 0 for $m \not\equiv a \bmod. 2$; for $m \equiv a \bmod. 2$, it is $\mathcal{G}(x^a\alpha)/\mathcal{G}(\alpha)$; in the latter case, it cannot be 0 unless α is a pole of $\mathcal{G}$, of the form x^{-a+2f} with $0 < 2f \leqq a$. Similarly, for $K = \mathbf{C}$ and $\Phi(x) = x^{a'}\bar{x}^{a''}e^{-2\pi x\bar{x}}$, $\Delta_\alpha(\Phi)$ is 0 unless α is of the form $x^{-a'}\bar{x}^{-a''}(x\bar{x})^s$; in the latter case, it is $\mathcal{G}((x\bar{x})^s)/\mathcal{G}(\alpha)$, i.e. $G_2(s)/\mathcal{G}(\alpha)$, and it is not 0 unless α is a pole of $\mathcal{G}$, s being an integer > 0 and $\leqq \inf(a', a'')$.

For $K = \mathbf{R}$, put $\beta(x) = x\alpha(x)$; then $\Delta_\alpha(x\Phi)$ is $(\zeta/2\pi)\Delta_\beta(\Phi)$ if $m = 0$, and $\Delta_\beta(\Phi)$ if $m = 1$. Similarly, for $K = \mathbf{C}$, put $\beta'(x) = x\alpha(x)$, $\beta''(x) = \bar{x}\alpha(x)$; then $\Delta_\alpha(x\Phi)$ is $(\zeta'/2\pi)\Delta_{\beta'}(\Phi)$ if $\zeta' - \zeta'' \geqq 0$, and $\Delta_{\beta'}(\Phi)$ otherwise; $\Delta_\alpha(\bar{x}\Phi)$ is $(\zeta''/2\pi)\Delta_{\beta''}(\Phi)$ if $\zeta' - \zeta'' \leqq 0$, and $\Delta_{\beta''}(\Phi)$ otherwise. On the other hand, for $K = \mathbf{R}$, put $\gamma(x) = x^{-1}\alpha(x)$; by combining (iii) with the results just mentioned (or by integrating by parts), we find that $\Delta_\alpha(\frac{d\Phi}{dx})$ is $(1 - \zeta)\Delta_\gamma(\Phi)$ if $m = 0$, and $-2\pi\Delta_\gamma(\Phi)$ if $m = 1$. For $K = \mathbf{C}$, put $\gamma'(x) = x^{-1}\alpha(x)$, $\gamma''(x) = \bar{x}^{-1}\alpha(x)$; then $\Delta_\alpha(\frac{\partial\Phi}{\partial x})$ is $(1 - \zeta')\Delta_{\gamma'}(\Phi)$ if $\zeta' - \zeta'' \leqq 0$ and $-2\pi\Delta_{\gamma'}(\Phi)$ if $\zeta' - \zeta'' > 0$; $\Delta_\alpha(\frac{\partial\Phi}{\partial \bar{x}})$ is $(1 - \zeta'')\Delta_{\gamma''}(\Phi)$ if $\zeta' - \zeta'' \geqq 0$ and $-2\pi\Delta_{\gamma''}(\Phi)$ if $\zeta' - \zeta'' < 0$.

By a <u>polynomial function</u> P on Ω_K, we will understand one which is a polynomial on each connected component of Ω_K, i.e. one such that $s \longrightarrow P(\omega_s\alpha)$ is a polynomial in s for all α. Such are e.g. $\alpha \longrightarrow \zeta$ for $K = \mathbf{R}$, $\alpha \longrightarrow \zeta'$ and $\alpha \longrightarrow \zeta''$ for $K = \mathbf{C}$. Using induction on n, one finds then, e.g. for $K = \mathbf{R}$, that $\Delta_\alpha(x^n\Phi) = P_n(\alpha)\Delta_{x^n\alpha}(\Phi)$,

with a polynomial function P_n; similarly, we have
$\Delta_\alpha\left(\frac{d^n\Phi}{dx^n}\right) = Q_n(\alpha)\Delta_{x^{-n}\alpha}(\Phi)$, with another such function Q_n; of course there are corresponding results for $K = \mathbf{C}$.

47. After these preliminaries, we can now prove one main portion of the "local functional equation" for $GL(2, \mathbf{R})$ and $GL(2, \mathbf{C})$.

Proposition 12. Let S be any Schwartz function on K^2; Ψ_S being defined by (31), put, for any $\omega \in \Omega_K$:

$$I(S, g, \omega) = \int_{K^\times} \Psi_S\left(\begin{pmatrix} u & 0 \\ 0 & 1 \end{pmatrix} . g\right)\omega(u)d^\times u \ ,$$

$$J(S, g, \omega) = \mathcal{G}(\omega)^{-1}\mathcal{G}(\omega_1\alpha\omega)^{-1}I(S, g, \omega) \ .$$

Then the integral $I(S, g, \omega)$ is absolutely convergent for $\sigma(\omega)$ large; $\omega \longrightarrow J(S, g, \omega)$ can be continued analytically to an entire function on Ω_K, and we have, for $j = \begin{pmatrix} 0 & 1 \\ -1 & 0 \end{pmatrix}$ and all S, g:

$$J(S, jg, \alpha^{-1}\omega^{-1}) = \kappa(\omega)\kappa(\alpha\omega)J(S, g, \omega) \ .$$

For brevity, put $\gamma_g = |\det g|_K^{-1}$; (31) can be written as

$$\Psi_S(g) = \gamma_g \int S'_g(u/y, y)\alpha(y)|y|_K d^\times y \ ,$$

with S'_g as defined in §44. Write $I(S, g, \omega)$ formally as a double integral, and replace u by uy; we get:

$$I(S, g, \omega) = \gamma_g \int S'_g(u, y)\omega(u).(\omega_1\omega\alpha)(y)d^\times u d^\times y \ .$$

This is absolutely convergent when $\sigma(\omega)$ and $\sigma(\omega_1\alpha\omega)$ are both > 0, so that the original integral for $I(S, g, \omega)$ must itself be so. With the notation of lemma 6, we can now write

$$J(S, g, \omega) = \gamma_g(\Delta_\omega \otimes \Delta_{\omega_1\alpha\omega})(S'_g) \ .$$

In view of Tate's lemma, this gives the analytic continuation of $J(S, g, \omega)$ as an entire function on Ω_K. As to the functional equation, we have

$$\Psi_S\left(\begin{pmatrix} v & 0 \\ 0 & 1 \end{pmatrix} . jg\right) = \gamma_g \int S''_g(x, v/x)\alpha(-x)|x|_K d^\times x \ ,$$

where S''_g is the "partial Fourier transform"

$$S''_g(x, v) = \int S_g(x, y)\psi(-yv)dy \ .$$

Then S'_g, S''_g are essentially the Fourier transforms of each other on K^2. We also introduce the Fourier transform of S itself on K^2:

$$S^*(u, v) = \int S\left(\begin{pmatrix} x \\ y \end{pmatrix}\right)\psi(ux + vy)dxdy$$

in terms of which we can at once express that of S_g:

$$\int S_g(x, y)\psi(ux + vy)dxdy = |\det g|_K S^*((u, v). g) \ ,$$

where $(u, v). g$ is understood in the sense of matrix multiplication. We will write S^*_g for the function given by

$$S^*_g(u, v) = S^*((u, v). g) \ ,$$

i.e. for $\gamma_g . (S_g)^*$.

Just as above in the case of $J(S, g, \omega)$, we get now:

$$J(S, jg, \alpha^{-1}\omega^{-1}) = \gamma_g \alpha(-1)(\Delta_{\omega_1\omega^{-1}} \otimes \Delta_{\alpha^{-1}\omega^{-1}})(S''_g) \ .$$

As S'_g, S''_g are the Fourier transforms of each other on K^2, we have now only to apply lemma 6(iii) to both factors of the space $K^2 = K \times K$ in order to get the functional equation as stated in our proposition. By applying it to one factor at a time, we get:

(34) $$J(S, g, \omega) = \gamma_g \kappa(\omega)(\Delta_{\omega_1\omega^{-1}} \otimes \Delta_{\omega_1 a\omega})(S_g)$$
$$= \kappa(a\omega)^{-1}(\Delta_\omega \otimes \Delta_{a^{-1}\omega^{-1}})(S_g^*) \ .$$

48. Implicit in the functional equation of proposition 12 are the "gamma factors" $\mathcal{G}(\omega)\,\mathcal{G}(\omega_1 a\omega)$, $\mathcal{G}(a^{-1}\omega^{-1})\,\mathcal{G}(\omega_1\omega^{-1})$; these may have common poles. Clearly $\mathcal{G}(\omega)$ and $\mathcal{G}(\omega_1\omega^{-1})$ have none, so that also $\mathcal{G}(\omega_1 a\omega)$ and $\mathcal{G}(a^{-1}\omega^{-1})$ have none; taking into account the conditions imposed on a by (28) resp. (29), one sees that $\mathcal{G}(\omega_1 a\omega)$ and $\mathcal{G}(\omega_1\omega^{-1})$ have none. There remains the possibility that $\mathcal{G}(\omega)$ and $\mathcal{G}(a^{-1}\omega^{-1})$ may have common poles; one sees at once that this will be so if and only if a^{-1} is a pole of $\mathcal{G}$, and in particular, as (29) will show, in the case of a discrete type for $K = \mathbf{R}$. We will need the following lemma, which is relevant to that case:

Lemma 7. <u>For</u> $K = \mathbf{R}$, <u>let</u> a <u>be given by</u> $a(x) = x^{\rho-1}$, <u>where</u> ρ <u>is an integer</u> $\geqq 0$. <u>For</u> $\omega(x) = (\operatorname{sgn} x)^m |x|^s$, <u>put</u>

$$D_a(\omega) = G_2(s+\rho),\ \eta_a(\omega) = i^{\rho+1}\kappa(\omega)\kappa(a\omega) \ .$$

<u>Then</u> $\omega \longrightarrow D_a(\omega)^{-1}$ <u>and</u> $\omega \longrightarrow D_a(a^{-1}\omega^{-1})^{-1}$ <u>are entire functions without common zeros,</u> $\omega \longrightarrow \eta_a(\omega)$ <u>is locally constant; we have</u>

$$\frac{D_a(a^{-1}\omega^{-1})}{\mathcal{G}(\omega_1\omega^{-1})\,\mathcal{G}(a^{-1}\omega^{-1})} = \eta_a(\omega)\frac{D_a(\omega)}{\mathcal{G}(\omega)\,\mathcal{G}(\omega_1 a\omega)}$$

<u>and the common value</u> $P_a(\omega)$ <u>of both sides is a polynomial function on</u> Ω_K. <u>For</u> $\rho = 0$, <u>we have</u> $P_a(\omega) = \pi$, $\eta_a(\omega) = 1$.

This can be routinely verified by using the most elementary properties of the gamma function. One may note that $\eta_a(\omega)$ is always ± 1.

Proposition 13. <u>There is a Schwartz function</u> S, <u>satisfying the condition in the latter part of proposition</u> 11 <u>and having the following</u>

property: (a) except when $K = R$ and $[\alpha, n]$ is a discrete type, there is no $\omega \in \Omega_K$ for which $J(S, g, \omega)$ is 0 for all $g \in G$; (b) for $K = \mathbf{R}$, if $[\alpha, n]$ is a discrete type, and P_α is the polynomial function defined in lemma 7, $\omega \longrightarrow P_\alpha(\omega)^{-1} J(S, g, \omega)$ is an entire function, and there is no ω for which $P_\alpha(\omega)^{-1} J(S, g, \omega)$ is 0 for all $g \in G$.

Clearly S will satisfy the condition in question if and only if its Fourier transform S^* satisfies $S^*((u, v)\mathfrak{k}) = S^*(u, v) M_n(\mathfrak{k})$ for all $\mathfrak{k} \in \mathfrak{K}$; it is easily seen that all such functions are to be obtained as follows. Let E be any function on $\mathbf{R}_+$ such that $E(t^2)$ is a Schwartz function on $\mathbf{R}$. For $K = \mathbf{R}$, we should take $S^*(u, v) = E(u^2 + v^2)$ if $n = 0$, and

$$S^*(u, v) = E(u^2 + v^2)((u + iv)^n, (-u + iv)^n)$$

for $n > 0$. If $K = \mathbf{C}$, notations being as in (28), we can define an integer h by putting $n - 2h = \rho' - \rho''$; then $0 \leqq h \leqq n$, and we should take $S^* = (S_o^*, \ldots, S_n^*)$ with

$$\sum_{j=0}^{n} S_j^*(u, v) X^{n-j} Y^j = E(u\bar{u} + v\bar{v})(uX + vY)^{n-h}(\bar{v}X - \bar{u}Y)^h \quad ;$$

in other words, the S_j^* are the coefficients of the right-hand side regarded as a polynomial in X, Y.

Just as in the proof of proposition 11, we observe now that the operators X, etc., resp. X', X'', etc., applied to $g \longrightarrow S^*((u, v).g)$, give functions $S_X^*((u, v).g)$, etc.; writing $X(S^*)$ instead of S_X^*, we have formulas, analogous to (33), viz. $X(S^*) = u \frac{\partial S^*}{\partial v}$, $Y(S^*) = v \frac{\partial S^*}{\partial u}$, etc., for $K = \mathbf{R}$, and $X'(S^*) = u \frac{\partial S^*}{\partial v}$, $Y'(S^*) = v \frac{\partial S^*}{\partial u}$, etc., for $K = \mathbf{C}$. Here we find that $g \longrightarrow S^*((u, v).g)$ is annulled by $D - \frac{1}{2} Z^2 - Z$ for $K = \mathbf{R}$, and conclude from the second formula (34) that $J(S, g, \omega)$ is harmonic of type (α, δ); as it obviously belongs to the representation M_n of $\mathfrak{K}$, it is therefore real-analytic on G (cf. §38); from its behavior under $g \longrightarrow gs$ it follows then that it must be 0 on G if all its derivatives at $g = 1_2$ are 0, i.e. if it is annulled by all distributions with the support $\{1_2\}$ on G. Exactly in the same way, we see, for $K = \mathbb{C}$, that $J(S, g, \omega)$ is harmonic of type $(\alpha, \delta', \delta'')$ and real-analytic, with the same conclusion as for $K = \mathbf{R}$.

Taking now $E(t) = e^{-\pi t}$ if $K = \mathbf{R}$, $E(t) = e^{-2\pi t}$ if $K = \mathbb{C}$, and using the second formula (34) and the results recalled in §46, one can calculate explicitly, not only $J(S, 1_2, \omega)$, but also $T[J(S, g, \omega)]$ if T is any distribution with the support $\{1_2\}$ on G: note that any such distribution can be written as a (non-commutative) polynomial in the operators X, etc., resp. X', X'', etc., at 1_2. To begin with, it is obvious that all these are polynomial functions of ω on Ω_K. Secondly, for $K = \mathbb{C}$, take $\omega(x) = \bar{x}^{n-h}(x\bar{x})^s$, and take the 0-th component of the (vector-valued) function $J(S, 1_2, \omega)$; we find for it the value $\alpha(-1)\kappa(\alpha\omega)$; as it is not 0, we conclude that Ψ_S is not 0 and is therefore standard; not only does this supply the missing existence proof for the standard function for $K = \mathbb{C}$, but it shows that it can be written in the form Ψ_S, so that we can apply to it the functional equation of proposition 12. For a similar conclusion in the case $K = \mathbf{R}$, one can write down the formula for $J(S, 1_2, \omega)$ in that case; this will be done presently, and it will be seen that it is not identically 0.

To prove the remainder of proposition 13, take first $K = \mathbb{C}$. Take first the 0-th component of $T[J(S, g, \omega)]$ for $T = X'^{\nu}$ at $g = 1_2$, and $\omega(x) = x^{h-n-\nu}(x\bar{x})^s$; one finds that it has a constant value $\neq 0$. One finds the same for the n-th component for $T = Y'^{\nu}$ and $\omega(x) = x^{h+\nu}(x\bar{x})^s$. Thus, if $J(S, g, \omega)$ is 0 for all g for some $\omega(x) = x^r(x\bar{x})^s$, we must have $h - n < r < h$. For such an ω, take the n-th component of $T[J(S, g, \omega)]$ for $T = X'^{h-r}$; we find that either ω

or $\omega_1 \alpha\omega$ must be a pole of $\mathcal{G}$. Take the 0-th component of $T[J(S, g, \omega)]$ for $T = Y'^{r+n-h}$; we find that $\omega_1\omega^{-1}$ or $\alpha^{-1}\omega^{-1}$ must be a pole of $\mathcal{G}$. Thus ω must be a common pole of $\mathcal{G}(\omega)\mathcal{G}(\omega_1\alpha\omega)$ and $\mathcal{G}(\alpha^{-1}\omega^{-1})\mathcal{G}(\omega_1\omega^{-1})$, hence, as we have seen, of $\mathcal{G}(\omega)$ and $\mathcal{G}(\alpha^{-1}\omega^{-1})$. With the notations of (28), this implies that ρ', ρ'' are integers $\geqq 1$, so that $\rho' + \rho'' > n$ by definition of a reduced type, and therefore $\rho' > n - h$, $\rho'' > h$, and also that ω is of the form $\omega(x) = x^{-f'}\bar{x}^{-f''}$ with integers f', f'', and $0 \leqq f' < \rho'$, $0 \leqq f'' < \rho''$. Then $\alpha^{-1}\omega^{-1}$ is $x \longrightarrow x^{-h'}\bar{x}^{-h''}$ with $h' = \rho' - 1 - f'$, $h'' = \rho'' - 1 - f''$, and (34), together with the results in §46, shows that, up to a constant factor $\neq 0$, $T[J(S, g, \omega)]$ is

$$\left(\frac{\partial^{\rho'+\rho''-2}T(S^*)}{\partial u^{f'}\partial \bar{u}^{f''}\partial v^{h'}\partial \bar{v}^{h''}}\right)_0 .$$

Taking $T = X'^{\mu}X''^{\nu}$ with $0 \leqq \mu \leqq f'$, $0 \leqq \nu \leqq f''$, we see that this is

$$(\partial^{\rho'+\rho''-2}S^*/\partial u^{f'-\mu}\partial \bar{u}^{f''-\nu}\partial v^{h'+\mu}\partial \bar{v}^{h''+\nu})_0$$

with a coefficient $\neq 0$; using similarly the operators $T = Y'^{\mu}X''^{\nu}$, $X'^{\mu}Y''^{\nu}$, $Y'^{\mu}Y''^{\nu}$, we see that these are not all 0 unless all derivatives of S^* at 0, of order $\rho' - 1$ with respect to (u, v) and $\rho'' - 1$ with respect to $(\bar{u}, \bar{v})$, are 0. Putting now

$$e = \rho' - 1 - (n - h) = \rho'' - 1 - h ,$$

we see that those derivatives, for our choice of E, are, up to a non-zero factor, those of

$$(u\bar{u} + v\bar{v})^e (uX + vY)^{n-h}(\bar{v}X - \bar{u}Y)^h ,$$

and it is now clear that they cannot all be 0. This completes our proof for $K = \mathbf{C}$.

Now take $K = \mathbf{R}$; let ω be given by $\omega(x) = (\operatorname{sgn} x)^m |x|^s$ with $m = 0$ or 1; then we have $(\alpha^{-1}\omega^{-1})(x) = (\operatorname{sgn} x)^{m'} |x|^{s'}$, with $s' = -s-\rho+1$, $m' = 0$ or 1, and $m' \equiv m+n \bmod. 2$. For $n = 0$, one sees at once that $J(S, 1_2, \omega)$ is a non-zero constant for $m = 0$, and that $X[J(S, g, \omega)]$, taken at $g = 1_2$ as always, is a non-zero constant for $m = 1$. Before treating the case $n > 0$, we make the following observation, valid for any S (not merely the one chosen above). For any integer $\nu \geqq 0$, put $\omega'_\nu(x) = x^{2\nu}\omega(x)$, $\omega''_\nu(x) = x^{-2\nu}\omega(x)$; then, as induction on ν will show, we have:

$$X^{2\nu}[J(S, g, \omega)] = (2\pi i)^{2\nu} \frac{\mathcal{G}(\omega'_\nu)\,\mathcal{G}(\omega_1 \alpha\omega'_\nu)}{\mathcal{G}(\omega)\,\mathcal{G}(\omega_1\alpha\omega)} \cdot J(S, 1_2, \omega'_\nu)$$

$$Y^{2\nu}[J(S, g, \omega)] = (2\pi i)^{2\nu} \frac{\mathcal{G}(\omega_1 {\omega''_\nu}^{-1})\,\mathcal{G}(\alpha^{-1}{\omega''_\nu}^{-1})}{\mathcal{G}(\omega_1\omega^{-1})\,\mathcal{G}(\alpha^{-1}\omega^{-1})} \cdot J(S, 1_2, \omega''_\nu) \;.$$

Here the coefficients of $J(S, 1_2, \omega'_\nu)$ resp. $J(S, 1_2, \omega''_\nu)$ are polynomial functions, whose zeros are all among the poles of the denominators. Now take S as chosen above, with $E(t) = e^{-\pi t}$; then one finds that both components of $J(S, 1_2, \omega)$, up to non-zero constant factors, are equal to the polynomial function

$$F(\omega) = F_m(s) = \sum_{j \equiv m(2)} \binom{n}{j} (s+m)(s+m+2)\ldots(s+j-2).$$
$$(s+\rho-m'-1)(s+\rho-m'-3)\ldots(s+\rho-n+j+1) \;,$$

which is clearly of degree $\frac{1}{2}(n-m-m')$ in s (the highest coefficient being 2^{n-1}), and therefore $\neq 0$ on both components of Ω_K, as we had announced. Replacing ω by ω'_ν resp. ω''_ν changes s into $s+2\nu$ resp. $s-2\nu$; taking now the above formulas into account, we see, for ν large enough, that any common zero of all the functions $X^{2\nu}[J(S, g, \omega)]$, $Y^{2\nu}[J(S, g, \omega)]$ must be a common pole of

$\mathcal{G}(\omega)\mathcal{G}(\omega_1 \alpha\omega)$ and $\mathcal{G}(\omega_1\omega^{-1})\mathcal{G}(\alpha_1^{-1}\omega^{-1})$, and therefore, as we have seen, of $\mathcal{G}(\omega)$ and $\mathcal{G}(\alpha^{-1}\omega^{-1})$. For this to happen, we must have $\rho \in \mathbf{Z}$, $\rho \geqq 1$, $\rho \equiv n+1 \bmod. 2$, and therefore $\alpha(x) = x^{\rho-1}$, and $\omega(x) = x^{-f}$, $0 \leqq f \leqq \rho - 1$. Then, up to a non-zero constant factor, $J(S, g, \omega)$ is given by

$$\left(\frac{d^{\rho-1}[S^*((u, v).g)]}{du^f dv^{\rho-1-f}}\right)_0 .$$

Here we may replace S^* by its terms of order $\rho - 1$; these are 0 if $\rho < n$; otherwise, up to a constant factor, they are $(u^2 + v^2)^e(\pm u + iv)^n$ with $e = (\rho - 1 - n)/2$ (note that $\rho \equiv n + 1 \bmod. 2$). In the former case, i.e. when the type $[\alpha, n]$ is discrete, we have $J(S, g, \omega) = 0$ for all g; in the latter case, i.e. when $[\alpha, n]$ is principal, this is not so, since no coefficient of a homogeneous polynomial in u, v can remain 0 under all substitutions $(u, v) \longrightarrow (u, v).g$. Finally, in the case of a discrete type, the above formula for $X^{2\nu}[J(S, g, \omega)]$ shows that ω is a simple zero of that function for ν large enough. In view of lemma 7, this completes the proof for $K = \mathbf{R}$.

Corollary. <u>Let</u> T <u>be any distribution with the support</u> $\{1_2\}$ <u>on</u> G. <u>Then</u> $T[J(S, g, \omega)]$ <u>is a polynomial function on</u> Ω_K, <u>and a multiple of</u> P_α <u>if</u> $K = \mathbf{R}$ <u>and</u> $[\alpha, n]$ <u>is a discrete type; moreover, the functions</u> $P_\alpha(\omega)^{-1}T[J(S, g, \omega)]$ <u>in the latter case, and in all other cases the functions</u> $T[J(S, g, \omega)]$, <u>have no common zero on</u> Ω_K.

49. In view of the observations in §43, it is obvious that the results in §§47-48 imply similar ones for arbitrary (non-reduced) types. This will now be made explicit. First of all, we describe the gamma factors and the constant factors in the functional equation in the general case:

(a) For $K = \mathbf{C}$, let a type $(\alpha, \delta', \delta'', n)$ be given; define λ as in §43, i.e. so that $(\alpha\lambda^2, \delta', \delta'', n)$ is the reduced type

$[\alpha\lambda^2, n]$. Then we put

$$(35) \qquad G(\omega) = \mathcal{G}(\lambda^{-1}\omega)\,\mathcal{G}(\omega_1\alpha\lambda\omega), \quad e(\omega) = \kappa(\lambda^{-1}\omega)\kappa(\alpha\lambda\omega) \ .$$

(b) For $K = \mathbf{R}$, let a principal type (α, δ, d, n) be given; define λ as in §43, i.e. so that $\lambda(-1) = (-1)^{n-d}$ and the $(\alpha\lambda^2, \delta, d', n)$ is the reduced type $[\alpha\lambda^2, n]$ for $d' \equiv n \bmod. 2$. Then we again define G, e by the formula (35).

(c) For $K = \mathbf{R}$, let a discrete type (α, δ, n) be given; define λ as in §43, i.e. so that $(\alpha\lambda^2, \delta, n)$ is the reduced type $[\alpha\lambda^2, n]$; this implies that $(\alpha\lambda^2)(x) = x^{\rho-1}$ and $\delta = (\rho^2 - 1)/2$, ρ being an integer ≥ 0. If then $(\lambda^{-1}\omega)(x) = (\operatorname{sgn} x)^m |x|^s$, we put

$$(36) \qquad G(\omega) = G_2(s + \rho), \quad e(\omega) = i^{-1-\rho} \ .$$

Now we summarize the main results of this Chapter as follows:

Theorem 4. (i) For every admissible type on G, there is one and, up to a constant factor, only one standard function Φ; for $p > 0$, the function $p \longrightarrow \Phi(p^{-1/2}\begin{pmatrix} p & 0 \\ 0 & 1 \end{pmatrix})$ is $O(e^{-Ap})$, with a suitable $A > 0$, for $p \longrightarrow +\infty$, and $O(p^{-B})$, with a suitable B, for $p \longrightarrow 0$; so are all its derivatives.

(ii) The integral

$$I(g, \omega) = \int_{K^\times} \Phi(\begin{pmatrix} u & 0 \\ 0 & 1 \end{pmatrix} . g)\omega(u)d^\times u$$

is absolutely convergent for $\sigma(\omega)$ large; it can be continued analytically to a meromorphic function on Ω_K.

(iii) Let the "gamma factor" $G(\omega)$ and the locally constant function $e(\omega)$ be defined on Ω_K by (35) resp. (36); put

$$J(g, \omega) = G(\omega)^{-1} I(g, \omega) \ .$$

Then $J(g, \omega)$ is an entire function of ω on Ω_K, satisfying the functional equation

$$J(jg, \alpha^{-1}\omega^{-1}) = e(\omega)J(g, \omega) \;;$$

for each ω $g \longrightarrow J(g, \omega)$ is real-analytic on G and not 0.

(iv) For each distribution T with the support $\{1_2\}$ on G, $T[J(g, \omega)]$ is a polynomial function on Ω_K; there is no common zero on Ω_K to all these functions.

Remark 1. As will be seen from (35), (36), G and e depend only upon ω, α and λ; as to λ, as defined in §43, it depends only upon $\alpha, \delta', \delta''$ in case (a), upon α, δ, d in case (b), α, δ in case (c); thus, in each case, it does not depend upon n, and all the choices of n (compatible with the conditions laid down for an admissible type in each case) lead to one and the same functional equation for $J(g, \omega)$. This finds its natural explanation in representation-theory, where it is shown that all these values of n correspond to one and the same infinite-dimensional representation of G, which is characterized by the function G and e. On the other hand, it also gives the possibility, for given G and e (i.e. for a given functional equation), of choosing n so as to make the formulas as simple as possible. We now indicate some such choices:

a) For $K = \mathbf{R}$ and a discrete type, it is natural to take for n the smallest permissible value $n = \rho + 1$. This choice, which will be discussed more fully (as type $\mathcal{H}_{n,\alpha}$) in the next Chapter, is the one underlying the classical theory of holomorphic automorphic forms of degree n.

b) For $K = \mathbf{R}$ and a principal type for which $\alpha(-1) = 1$, the simplest choice is given by $n = 0$, in which case the standard function Φ is given by Hankel's function; this is the choice underlying Maass's pioneering work on non-holomorphic automorphic functions in the upper half-plane. For $\alpha(-1) = -1$, one may of course take $n = 1$, but this does not seem to have any advantage over other choices.

c) For $K = \mathbb{C}$, the simplest choice for n is the smallest value compatible with the other data; this is $n = |\rho' - \rho''|$, except when $\rho' = (1 + 2\delta')^{1/2}$ and $\rho'' = (1 + 2\delta'')^{1/2}$ are both integers and not 0; taking then $\rho' > 0$, $\rho'' > 0$, one has to distinguish two cases, according as $n \geqq \rho' + \rho''$ or $|\rho' - \rho''| \leqq n < \rho' + \rho''$; in each one of these cases (which lead to different functional equations) one will take for n the smallest value allowed by these inequalities. The "special case" $\mathcal{H}_{\mathbb{C}}$ to be discussed in the next Chapter is given by $\delta' = \delta'' = 0$, $\rho' = \rho'' = 1$, $n = 2$.

Remark 2. In §41, Remark 2, we have already observed that virtually no changes are required in our definitions in the case $K = \mathbb{C}$ if, instead of one of the representations M_n, we make use of one which is merely equivalent to M_n; if $M = A^{-1}.M_n.A$, and if Φ is the standard function belonging to M_n, ΦA is the standard function belonging to M; clearly theorem 4 applies equally well to it, the functional equation remaining the same.

Remark 3. In Chapter IV, we introduced the concept of a "dual pair"; this has played an important role in Chapter VII; as it will again do so in Chapter X, we must make some remarks here on the operation $\Phi \longrightarrow \Phi'$ defining that concept (as given by (7), §17) and its effect on admissible functions. At an infinite place, it is given by

$$\Phi'(g) = \Phi(g\mathfrak{a})\alpha(\det g)^{-1}$$

with $\mathfrak{a} = \begin{pmatrix} 0 & 1 \\ -1 & 0 \end{pmatrix}$. For $K = \mathbb{C}$, let Φ be admissible of type $(\alpha, \delta', \delta'', M)$ in the sense of Remark 2, §41, and of Remark 2 above, with M equivalent to M_n; then Φ' is admissible of type $(\alpha^{-1}, \delta', \delta'', M')$ with $M'(\mathfrak{p}) = M(\mathfrak{a}^{-1}\mathfrak{p}\mathfrak{a})$; if one is standard, so is the other. For $K = \mathbb{R}$, we have $\mathfrak{a} = r(\pi/2)$; it is easily verified that Φ' belongs to the same representation (M_n, M_o or M'_o) as Φ; moreover, the inner automorphism $g \longrightarrow a^{-1}ga$, and therefore also the right-translation $g \longrightarrow ga$, change the operator W into $-W$, and $\overline{W}$ into $-\overline{W}$. From this, one concludes at once that, if Φ is admissible for the principal type (α, δ, d, n), Φ' is so for the type $(\alpha^{-1}, \delta, d', n)$ with $d' \equiv d + n \bmod. 2$, and that, if it is admissible for the discrete type (α, δ, n), Φ' is so for the type (α^{-1}, δ, n); if one is standard, so is the other.

CHAPTER IX

HARMONICITY (SPECIAL CASE)

50. Except for brief references to some of the definitions of Chapter VIII, the present Chapter will be independent of the latter; the basic notations will be those explained there in §§33-35.

As we have noted in §34, the subgroup B_1 of G provides us with a complete set of representatives for the cosets $g\mathfrak{K}\mathfrak{Z}$ and may be identified with the Riemannian symmetric spaces $H = G/\mathfrak{K}\mathfrak{Z}$. We will write π for the canonical mapping (the "projection") of G onto $H = B_1$; thus, every element g of G may be written as $g = b_1\mathfrak{k}\mathfrak{z}$, with $b_1 = p^{-1/2}\begin{pmatrix} p & y \\ 0 & 1 \end{pmatrix}$, $p > 0$, $y \in K$, $\mathfrak{k} \in \mathfrak{K}$ (and even $\mathfrak{k} \in \mathfrak{K}_1$ in the case $K = \mathbf{C}$), $\mathfrak{z} \in \mathfrak{Z}$, and we write then $b_1 = \pi(g)$. The group G operates on H by $(g, b_1) \longrightarrow \pi(gb_1)$; the invariant Riemannian structure in H is given by $ds^2 = p^{-2}(dp^2 + dyd\bar{y})$.

We wish to consider harmonic differential forms in H, and their inverse images (or "pullbacks") in G. It will be convenient to express them in terms of those left-invariant differential forms on G which are 0 on $\mathfrak{K}\mathfrak{Z}$; a basis (ω_i) for the latter will now be chosen as follows. On the group B_1, we take a basis (β_i) for the left-invariant differential forms, given by

$$\beta_1 = p^{-1}(dy + idp),\ \beta_2 = p^{-1}(-dy + idp) \qquad \text{for } K = \mathbf{R}\ ,$$

$$\beta_0 = -p^{-1}dy,\ \beta_1 = p^{-1}dp,\ \beta_2 = p^{-1}d\bar{y} \qquad \text{for } K = \mathbf{C}\ .$$

Call ω_i the left-invariant differential form on G which, at the point 1_2, coincides with the inverse image $\pi^*\beta_i$ of β_i in G; it induces β_i on B_1. Right-translations defined by elements of $\mathfrak{K}\mathfrak{Z}$ in G operate on the ω_i through a representation of $\mathfrak{K}\mathfrak{Z}$ which is obviously trivial on $\mathfrak{Z}$. Writing $\omega = \begin{pmatrix}\omega_1\\ \omega_2\end{pmatrix}$ resp. $\omega = \begin{pmatrix}\omega_o\\ \omega_1\\ \omega_2\end{pmatrix}$ as a column-vector, we can write this representation as $\omega \longrightarrow M(\mathfrak{p}\mathfrak{z})^{-1}\omega$; a simple calculation shows that, if $K = \mathbf{R}$, M coincides on $\mathfrak{K}$ with the representation M_2 of $\mathfrak{K}$ defined in §35, a), and, if $K = \mathbf{C}$, it coincides on $\mathfrak{K}_1$ with the representation M_2 of $\mathfrak{K}_1$ defined in §35, b); of course this would not have been so if the basis ω_i had not been suitably chosen. Now write any g as $g = b_1\mathfrak{p}\mathfrak{z}$, with $b_1 = \pi(g)$, $\mathfrak{p} \in \mathfrak{K}$ (if $K = \mathbf{R}$) and even $\mathfrak{p} \in \mathfrak{K}_1$ (if $K = \mathbf{C}$), and $\mathfrak{z} \in \mathfrak{Z}$; it is clear that we have then, in an obvious sense, $\omega(g) = M_2(\mathfrak{p})^{-1}\pi^*\beta(g)$. A differential form on G is then the inverse image of one on H if and only if it can be written as $\Sigma\,\varphi_i\omega_i$, where the vector-valued function $\Phi = (\varphi_1, \varphi_2)$ resp. $\Phi = (\varphi_o, \varphi_1, \varphi_2)$ satisfies $\Phi(g\mathfrak{p}\mathfrak{z}) = \Phi(g)M(\mathfrak{p}\mathfrak{z})$ for all $g \in G$ and all $\mathfrak{p}\mathfrak{z}$ in $\mathfrak{K}\mathfrak{Z}$, M being as we have said above; as in §35, we express this by saying that Φ belongs to the trivial character $\alpha = 1$ of $\mathfrak{Z}$ and to the representation M_2 of $\mathfrak{K}$ resp. $\mathfrak{K}_1$.

51. Now we wish to express that such a form $\Sigma\,\varphi_i\omega_i$ is the inverse image of a harmonic form on H, for the invariant Riemannian structure defined above. Clearly that is so if and only if the form $\Sigma\,f_i\beta_i$ induced on B_1 is harmonic, f_i being the function induced by φ_i on B_1.

Take first the case $K = \mathbf{R}$; then, as Φ belongs to M_2, we have for $s = \begin{pmatrix}-1 & 0\\ 0 & 1\end{pmatrix}$, $\varphi_2(g) = \varphi_1(gs)$ for all g, and $\varphi_1(g.r(\theta)) = \varphi_1(g)e^{2i\theta}$, with $r(\theta)$ as in §35, a). Let $*$ be the usual operator on differential forms, as defined in Riemannian geometry, for the given ds^2; as $ds^2 = \beta_1\overline{\beta}_1$ and $\beta_2 = -\overline{\beta}_1$, we have $*(f_1\beta_1 + f_2\beta_2) = i^{-1}(\overline{f}_1\beta_2 - \overline{f}_2\beta_1)$; thus the given form is harmonic if it is closed and the form $\overline{f}_1\beta_2 - \overline{f}_2\beta_1$, or, what amounts to the same, the form $f_1\beta_1 - f_2\beta_2$, is closed; this is the same as to say that both $f_1\beta_1$

and $f_2\beta_2$ are closed, or again, putting $\tau = y + ip$, that $f_1 p^{-1} d\tau$ and $\bar{f}_2 p^{-1} d\bar{\tau}$ are holomorphic differentials in the half-plane $p > 0$. If that is so, and if at the same time Φ is B-moderate in the sense of §33, we will say that Φ, and the differential form $\varphi_1\omega_1 + \varphi_2\omega_2$, are <u>admissible of type</u> $\mathcal{H}_{\mathbf{R}}$. We recall that Φ is B-moderate if and only if $f_1(p, y)$ and $f_2(p, y)$ are $O(p^N)$ for some N, uniformly in y over every compact interval in $\mathbf{R}$.

For $K = \mathbf{C}$, we have

$$*(\Sigma f_i\beta_i) = -\frac{i}{2}\bar{f}_1\beta_0 \wedge \beta_2 + i\bar{f}_0\beta_1 \wedge \beta_2 + i\bar{f}_2\beta_0 \wedge \beta_1 \ .$$

If this, together with the form $\Sigma f_i\beta_i$, is closed, the form is said to be harmonic; if at the same time Φ is B-moderate, we say that Φ and the form $\Sigma \varphi_i\omega_i$ are <u>admissible of type</u> $\mathcal{H}_{\mathbf{C}}$.

The function Φ, and the form $\Sigma \varphi_i\omega_i$, will be called standard if they are admissible of type $\mathcal{H}_K$, not 0, and satisfy the condition

$$\Phi\left(\begin{pmatrix} 1 & y \\ 0 & 1 \end{pmatrix} \cdot g\right) = \psi(y)\Phi(g) \tag{37}$$

for all $y \in K$ and all $g \in G$. Clearly this is so if and only if the functions f_i all have that property. In the case $K = \mathbf{R}$, $f_1 p^{-1}$ and $\bar{f}_2 p^{-1}$ must be holomorphic, so that this gives $f_1 = C_1 pe^{-2\pi i\tau}$, $f_2 = C_2 pe^{-2\pi i\bar{\tau}}$, with constants C_1, C_2; as they must be B-moderate, C_1 must be 0. Therefore, up to a constant, there is one and only one standard function (or form).

In the case $K = \mathbf{C}$, writing $f_i(p, y) = f_i(p)e^{-2\pi i(y+\bar{y})}$, and expressing that $\Sigma f_i\beta_i$ is closed, we get $f_2 = -f_0$ and $\frac{d}{dp}(p^{-1}f_0) = 2\pi ip^{-1}f_1$. Expressing that $*(\Sigma f_i\beta_i)$ is closed, we get $\frac{d}{dp}(p^{-2}f_1) = 4\pi ip^{-2}(f_2 - f_0)$. Writing $z = 4\pi p$ and $p^{-2}f_1(p) = K(z)$, we get at once for K the equation $zK'' + K' - zK = 0$, which is the classical equation for Hankel's function K_o; it is known (cf. e.g. W. Magnus, etc., loc. cit.) that K_o is the only solution of that equation which does not increase exponentially for $z \longrightarrow +\infty$. Consequently, in order that Φ may be standard, we must take $f_1(p) = p^2 K_o(4\pi p)$; then f_o and f_2 are given by the above formulas. As Hankel's function K_1 is given by $K_1(z) = -dK_o/dz$, we get $f_o(p) = -f_2(p) = -\frac{i}{2} p^2 K_1(4\pi p)$.

In both cases one verifies at once the validity of assertion (i) in theorem 4, §49.

52. For a comparison with the results of Chapter VIII, we insert here the following observations. Take first the case $K = \mathbf{R}$; let notations be as above, and call W (as in Chapter I, §4, and in Chapter VIII, §41) the left-invariant differential operator on G, defined by the element $\begin{pmatrix} 1 & -i \\ -i & -1 \end{pmatrix}$ of the Lie algebra. As we noted already in Chapter I, §4, and as may again be readily verified, $f_1\beta_1$, and consequently its inverse image $\varphi_1\omega_1$ in G, are closed if and only if $W\varphi_1 = 0$ (cf. also the formulas in the proof of proposition 10, §42); by formula (26) of §41, this gives $D\varphi_1 = 0$, and therefore $D\varphi_2 = 0$ since $\varphi_2(g) = \varphi_1(gs)$. Consequently, a function Φ is admissible of type $\mathcal{H}_{\mathbf{R}}$, according to the definition of §51, if and only if it is admissible of type (1, 0, 2) in the sense of Chapter VIII. This is a discrete type, and a reduced one (viz., the reduced type [1, 2]) in the sense of §43. The formulas found above for the standard function of that type agree of course with those of §42.

Now take $K = \mathbf{C}$; in that case, a somewhat laborious calculation, based on formula (24) of §37, shows that, if the differential form $\Sigma\, \varphi_i\omega_i$ is harmonic, then $D'\Phi = D''\Phi = 0$, and conversely, if $D'\Phi = D''\Phi = 0$, there is a constant C such that the form $(\Sigma\, f_i\beta_i) - C\beta_1$, and consequently its inverse image $(\Sigma\, \varphi_i\omega_i) - C\pi^*(\beta_1)$, are harmonic. Thus the space of admissible functions of type (1, 0, 0, 2) in the sense of §41 consists of the admissible functions of type $\mathcal{H}_{\mathbb{C}}$ and of one more function, corresponding to the form $\pi^*(\beta_1)$; one finds that the latter can be written as

$$\Phi(\begin{pmatrix} a & b \\ c & d \end{pmatrix}) = (c\bar{c} + d\bar{d})^{-1}(c\bar{d},\ d\bar{d} - c\bar{c},\ -d\bar{c}) \ .$$

So far as the definition of the standard function is concerned, there is thus no difference between the type $(1, 0, 0, 2)$ and the type $\mathcal{H}_{\mathbb{C}}$. One should note that $(1, 0, 0, 2)$ is not a reduced type in the sense of §43; with the notations of (28), we have, for that type, $\rho' = \rho'' = 1$, $n = 2$; the corresponding reduced type would be given by $\alpha(x) = x^{-2}$, $n = 2$, $\rho' = -1$, $\rho'' = 1$.

53. As appears from Chapter I, §4, the case discussed above for $K = \mathbf{R}$ corresponds to the case of the modular forms of degree 2 in the classical theory. In order to include modular forms of any degree, we proceed as follows. For any $n \geqq 1$, let α be a quasi-character of $\mathbf{R}^\times$ such that $\alpha(-1) = (-1)^n$. We consider functions φ_1 on G, satisfying $\varphi_1(g\mathfrak{z}) = \varphi_1(g)\alpha(\mathfrak{z})$ for all $\mathfrak{z} = z.1_2$ in $\mathfrak{Z}$, and $\varphi_1(g.r(\theta)) = \varphi_1(g)e^{in\theta}$; we also define φ_2 by $\varphi_2(g) = \varphi_1(gs)$ for all g. We say that φ_1, and the pair (φ_1, φ_2), are <u>admissible of type</u> $\mathcal{H}_{n,\alpha}$ if φ_1, φ_2 are B-moderate and if they induce on B_1 functions f_1, f_2 of the form $p^{n/2}F_1(\tau)$, $p^{n/2}\overline{F_2(\tau)}$, where F_1, F_2 are holomorphic in the half-plane $p > 0$. Formally, this can be expressed by saying that $f_1\beta_1^{n/2} = F_1.(d\tau)^{n/2}$ and $f_2\beta_2^{n/2} = \overline{F}_2.(d\bar{\tau})^{n/2}$ are holomorphic and antiholomorphic "differentials of degree $n/2$", respectively. As the same can also be expressed by $W\varphi_1 = 0$, $\overline{W}\varphi_2 = 0$, we see that this is the same as the concept of an admissible function of type $(\alpha, \frac{n^2}{2} - n, n)$ in the sense of §41; this is a discrete type; it is reduced if $\alpha(x) = x^{n-2}$. The standard function is again defined as the admissible function satisfying (37); one sees at once that it is given by $f_1 = 0$, $f_2 = p^{n/2}e^{-2\pi i\bar{\tau}}$. The type $\mathcal{H}_{\mathbf{R}}$ is of course

the special case $n = 2$, $\alpha = 1$ of the type $\mathcal{H}_{n,\alpha}$.

54. The "local functional equations" for the standard functions (which are special cases of those obtained in Chapter VIII) will now be restated in full.

For the topology and complex structure on the group Ω_K of the quasicharacters of $K^\times$, we refer to the beginning of §46. On this, we introduce, according to the type to be considered, a "gamma factor" $G(\omega)$ and a locally constant function $e(\omega)$, as follows:

a) Type $\mathcal{H}_{\mathbf{R}}$: for $\omega(x) = (\operatorname{sgn} x)^m |x|^s$, we put $G(\omega) = G_2(s+1)$, $e(\omega) = -1$; G_2 is the function defined in §46.

b) Type $\mathcal{H}_{n,\alpha}$: for $\alpha(x) = (\operatorname{sgn} x)^n |x|^\zeta$, and ω as in a), we put $G(\omega) = G_2(s + (n+\zeta)/2)$, $e(\omega) = i^{-n}$. As the type $\mathcal{H}_{\mathbf{R}}$ is no other than $\mathcal{H}_{2,1}$, we get a) again as a special case.

c) Type $\mathcal{H}_{\mathbf{C}}$: for $\omega(x) = x^{s'}\bar{x}^{s''}$, $s = \sup(s', s'')$, we put $G(\omega) = G_2(s+1)^2$, $e(\omega) = -1$ if $s' = s''$, and $G(\omega) = G_2(s)G_2(s+1)$, $e(\omega) = (-1)^{s'-s''}$ if $s' \neq s''$.

Theorem 5. (i) For each one of the types $\mathcal{H}_{\mathbf{R}}$, $\mathcal{H}_{n,\alpha}$, $\mathcal{H}_{\mathbf{C}}$, there is one and, up to a constant factor, only one standard function Φ; for $p > 0$, the function $p \longrightarrow \Phi(p^{-1/2}\begin{pmatrix} p & 0 \\ 0 & 1 \end{pmatrix})$ is $O(e^{-Ap})$, with a suitable $A > 0$, for $p \longrightarrow +\infty$, and $O(p^{-B})$, with a suitable B, for $p \longrightarrow 0$; so are all its derivatives.

(ii) The integral

$$I(g, \omega) = \int_{K^\times} \Phi\left(\begin{pmatrix} u & 0 \\ 0 & 1 \end{pmatrix} . g\right)\omega(u)d^\times u$$

is absolutely convergent for $\sigma(\omega)$ large; it can be continued analytically to a meromorphic function on Ω_K.

(iii) For $G(\omega)$, $e(\omega)$ defined as in a), b), c) above, the function $J(g, \omega) = G(\omega)^{-1}I(g, \omega)$ is an entire function of ω on Ω_K, satisfying

<u>the functional equation</u>

$$J(jg, \alpha^{-1}\omega^{-1}) = e(\omega)J(g, \omega).$$

(iv) <u>If the type is</u> $\mathcal{H}_R$ <u>or</u> $\mathcal{H}_{n,\alpha}$, <u>both components of</u> $J(1_2, \omega)$ <u>have constant non-zero values on each one of the two connected components of</u> Ω_K. <u>In all three cases, for each</u> ω, $g \longrightarrow J(g, \omega)$ <u>is real-analytic on</u> G <u>and not</u> 0; <u>for each distribution</u> T <u>with the support</u> $\{1_2\}$ <u>on</u> G, $T[J(g, \omega)]$ <u>is a polynomial function on</u> Ω_K, <u>and there is no common zero to all these functions.</u>

For $\mathcal{H}_R$, and more generally for $\mathcal{H}_{n,\alpha}$, this is easily verified by a straightforward elementary calculation (as was implicitly done by Hecke in his work on classical modular forms); this gives the additional piece of information given in (iv), although this could have been already obtained in Chapter VIII, had we wanted it at the time.

To verify it for the type $\mathcal{H}_{\mathbb{C}}$ (if one does not want simply to appeal to our Chapter VIII, i.e. essentially to Jacquet-Langlands) amounts to a rather elaborate exercise on Hankel's function and the hypergeometric function. Alternatively, one may in that case verify that the standard function obtained above can be written in the form (31) of §44, for a suitable choice of S. Actually one has to take, for j = 0, 1, 2:

$$S_j\left(\begin{pmatrix} x \\ y \end{pmatrix}\right) = (8\pi i)^{-1} e^{-2\pi(x\bar{x} + y\bar{y})} x^j (-y)^{2-j} .$$

Then, by using classical formulas (cf. again W. Magnus, etc., <u>loc. cit.</u>), we get, for $S = (S_o, S_1, S_2)$:

$$\Phi(g) = (\det \bar{g})^{-1} \int [S(g^{-1}\begin{pmatrix} x \\ y \end{pmatrix})\psi(\tfrac{x}{y})dx]y^{-2}dy ,$$

where $\bar{g}$ is the imaginary conjugate of g. Once this is obtained, the functional equation can be obtained just as in §47, and the results in (iv) as in §48.

CHAPTER X

NUMBER-FIELDS

55. Now we go back to global questions and to the notations of Chapters III and IV. From now on, k will be a number-field. We write n for its degree $[k : \mathbb{Q}]$ over $\mathbb{Q}$.

Let Φ be a function on G_A; as before, call F the function induced by Φ on B_A. We will say that Φ is B-*moderate* if there are constants $\lambda \geqq 0$, $C > 0$, such that

$$|F(x, y)| \leqq C \sup(|x|^{\lambda}, |x|^{-\lambda}) \tag{38}$$

for all $x \in k_A^\times$, $y \in k_A$.

Lemma 8. *Let Φ be a B-moderate function on G_A, satisfying conditions (A) to (D) of §12, and (E) of §14. Then there is $\mu \geqq 0$, and to every compact subset K of G_A, there is $C_K > 0$ such that*

$$\left|\Phi\left(\begin{pmatrix} p & 0 \\ 0 & 1 \end{pmatrix} \cdot g\right)\right| \leqq C_K |p|^{\mu}$$

for all $g \in K$ and for all $p = (p_v) \in k_A^\times$ such that $p_v = 1$ for v finite, $p_w \in \mathbb{R}$ and $p_w \geqq 1$ for w infinite.

It is clearly enough to show that, if μ is suitably chosen, every $g \in G_A$ has a compact neighborhood K with the above property. Moreover, as we have observed in §12, Remark 1, conditions (C) and (D) on Φ imply that Φ is locally constant with respect to the coordinates g_v of g at the finite places, so that it is enough if we prove our assertion when those coordinates are kept constant. Assume first that $g_v \in B_v \mathfrak{K}_v \mathfrak{Z}_v$ at the finite places which occur in the "conductor" $\mathfrak{A}$, and consequently at all places of k; write then $g = b\,\mathfrak{k}\,\mathfrak{z}$, with $b = (x, y) \in B_A$, $\mathfrak{k} \in \mathfrak{K}$, $\mathfrak{z} \in \mathfrak{Z}_A$; if g_∞ lies within a compact set in G_∞, so do x_∞ and $\mathfrak{z}_\infty$ in $k_\infty^\times$. In view of conditions (B), (C), (D), (E), the factors $\mathfrak{p}$, $\mathfrak{z}$ contribute only bounded factors to the value of Φ; this value can thus be estimated by (38); our assertion follows from this at once, with $\mu = \lambda$ and a suitable C_K. Now assume that, for some place v in $\mathfrak{A}$, g_v is not in $B_v \mathfrak{K}_v \mathfrak{z}_v$; then, as we have already observed at the beginning of §15, we can choose $\sigma = \begin{pmatrix} \alpha & \beta \\ \gamma & \delta \end{pmatrix}$ in G_k so that σg_v is in $B_v \mathfrak{K}_v \mathfrak{z}_v$ for all places v in $\mathfrak{A}$, and consequently for all places of k; in view of our assumption on g_v, we have $\gamma \neq 0$. Put $g' = \sigma g = b \mathfrak{p} \mathfrak{z}$, $b = (x, y)$. We have:

$$\Phi(\begin{pmatrix} p & 0 \\ 0 & 1 \end{pmatrix} \cdot g) = \Phi(\sigma \begin{pmatrix} p & 0 \\ 0 & 1 \end{pmatrix} \sigma^{-1} g') \ ;$$

just as before, if the coordinates g_v at the finite places are kept constant and g_∞ lies within a compact set, x_∞ and $\mathfrak{z}_\infty$ lie within compact subsets of $k_\infty^\times$, and $\mathfrak{p}$, $\mathfrak{z}$ contribute only bounded factors to the value of Φ; therefore it is enough to estimate the value of Φ at $g'' = \sigma \begin{pmatrix} p & 0 \\ 0 & 1 \end{pmatrix} \cdot \sigma^{-1} b$. This can be written (uniquely) as $b' \mathfrak{p}_\infty \mathfrak{z}_\infty$ with $b' = (x', y') \in B_A$, $b'_v = b_v$ for all finite v, $\mathfrak{p}_\infty \in \mathfrak{K}_\infty$, $\mathfrak{z}_\infty \in \mathfrak{Z}_\infty$, $\det(\mathfrak{p}_w) = 1$ and $\mathfrak{z}_w > 0$ for all infinite w. As the determinant of g''_w is $p_w x_w$, we have $x'_w = p_w x_w \mathfrak{z}_w^{-2}$. Assume first that $\delta = 0$. Then it is easily seen that $\mathfrak{z}_w = p_w$, hence $x'_w = p_w^{-1} x_w$; (38) shows that $\Phi(b')$ is $O(|p|^\lambda)$; $\mathfrak{p}_\infty$ contributes to $\Phi(g'')$ a bounded factor, while $\mathfrak{z}_\infty$ contributes the factor $\alpha_\infty(\mathfrak{z}_\infty)$, which is $O(|p|^{\lambda'})$ for a suitable λ'. Finally take the case $\delta \neq 0$; a simple calculation gives, for each infinite place w:

$$\mathfrak{z}_w^2 = q_w \bar{q}_w + r_w \bar{r}_w$$

where q_w, r_w are defined by

$$p'_w = (\det \sigma)^{-1}(p_w - 1) ,$$

$$q_w = 1 + \gamma(\delta y_w - \beta)p'_w , \quad r_w = \gamma\delta x_w p'_w ;$$

as usual, $\bar{q}_w = q_w$ and $\bar{r}_w = r_w$ if $k_w = \mathbf{R}$. Here x_w lies in a compact subset of $k_w^\times$, and y_w in a compact subset of k_w; from this it follows at once that $|p_w^{-1} \mathfrak{z}_w|$ lies in some interval $[A, B]$ with $0 < A < B$; one can then proceed just as in the case $\delta = 0$, with the same conclusion. This completes the proof of the lemma.

In particular, the lemma shows that, if Φ is B-moderate on G_A, the function it induces on G_w for each infinite place w (all coordinates of g, other than g_w, being kept constant) is B-moderate on G_w in the sense of §33.

56. As in Chapters III and IV, we assume that we have chosen a divisor $\mathfrak{a}$, a quasicharacter α of $k_A^\times/k^\times$, and an irreducible representation M_∞ of $\mathfrak{K}_\infty$ (compatible with α, i.e. agreeing with α_∞ on the center of $\mathfrak{K}_\infty$). As observed in §14, Remark 1, M_∞ is the tensor-product $\otimes M_w$ of irreducible representations M_w of the groups $\mathfrak{K}_w$; replacing these, if necessary, by equivalent ones, we may assume that they are of the form required for the application of the results of Chapter VIII. In view of Remark 3, §49, this puts no restriction on M_w if $k_w = \mathbf{C}$; if $k_w = \mathbf{R}$, it implies that M_w is one of the representations M_o, M'_o, M_n of §35, a); then we write $n_w = 0$ if M_w is M_o or M'_o, in which case its representation-space V_w is $\mathbf{C}$, and $n_w = n$ if M_w is M_n with $n > 0$, in which case $V_w = \mathbf{C}^2$; on the other hand, for $k_w = \mathbf{C}$, we write $n_w = n$ if m_w induces on $\mathfrak{K}_1$ a representation equivalent to M_n, in which case its representation-space V_w has the dimension $n_w + 1$ over $\mathbf{C}$. The representation-space of M_∞ is then $V = \otimes V_w$.

For each infinite place w, we choose now, in addition to α_w

and M_w, the data required for the determination of an admissible type on G_w in the sense of Chapter VIII, §41. Such data are, for $k_w = \mathbf{C}$, the two eigenvalues δ'_w, δ''_w for the Casimir operators D'_w, D''_w on G_w, subject to the conditions stated in proposition 9, §40. For $k_w = \mathbf{R}$, those data consist of an eigenvalue δ_w for the Casimir operator D_w, and in addition, as explained in §41, an integer d_w, equal to 0 or 1, unless $(\alpha_w, \delta_w, n_w)$ is a "discrete type". Then we say that all these data define an admissible type on G_∞, inducing on the factors G_w of G_∞ the types which make up its definition.

Alternatively, if $\alpha_\infty = 1$ and $n_w = 2$ for all w, we may assign to each place w the type $\mathcal{H}_w$ in the sense of Chapter IX, i.e. $\mathcal{H}_{\mathbf{R}}$ or $\mathcal{H}_{\mathbf{C}}$ according as k_w is R or $\mathbf{C}$; then we say that these define on G_∞ the type $\mathcal{H}_\infty$.

57. A V-valued function Φ on G_∞ will be called <u>admissible</u> of a given type if, for each w, it is admissible as a function of g_w (all other coordinates being kept constant) of the type induced by it on G_w. In so far as this involves only the operation on Φ of the Casimir operators, no further explanation is required here, since it is understood that these operate componentwise on vector-valued functions. As to the additional condition (involving the operator W) on admissible functions for $k_w = \mathbf{R}$, $n_w > 0$, this has to be understood as follows. Write $V = V' \otimes V_w$, where $V_w = \mathbf{C}^2$ and V' is the tensor-product of the spaces V_{w_1} for $w_1 \neq w$; we can write $\Phi = (\varphi_1, \varphi_2)$, where φ_1, φ_2 are V'-valued; then the operators $W, \overline{W}$ for the place w should be understood to operate componentwise on φ_1 and φ_2, and the condition in question should be interpreted accordingly.

As to the type $\mathcal{H}_\infty$, admissible functions for that type may be defined as follows. As in Chapter IX, §50, we introduce for each w the Riemannian symmetric space H_w belonging to G_w; then the product

$H_\infty = \prod H_w$ is the Riemannian symmetric space for G_∞. For each w, let the differential forms β_i on H_w, and ω_i on G_w, be as defined in §50; we may look upon (β_1, β_2) resp. $(\beta_0, \beta_1, \beta_2)$ as a V_w-valued differential form β_w on H_w, and similarly upon (ω_1, ω_2) resp. $(\omega_0, \omega_1, \omega_2)$ as a V_w-valued differential form ω_w on G_w; if r is the number of infinite places of k, we may, in an obvious sense, regard $\otimes\beta_w$, $\otimes\omega_w$ as V-valued differential forms of degree r on H_∞ and on G_∞. Then Φ is admissible of type $\mathcal{H}_\infty$ if and only if it is B-moderate on each G_w and the differential form $\Phi.\omega$ of degree r (the scalar product being understood in the obvious sense) is the inverse image on G_∞ of a harmonic form Ω on H_∞. One will note that the differential form $*\Omega$ (the $*$ being again understood in the sense of Riemannian geometry) is then of degree $[k : \mathbb{Q}]$.

58. A standard function of any type on G_∞ will be a non-zero admissible function Φ of that type, satisfying

$$\Phi(\begin{pmatrix} 1 & y \\ 0 & 1 \end{pmatrix}.g) = \psi_\infty(y)\Phi(g)$$

for all $y \in k_\infty$ and all $g \in G_\infty$. In Chapter VIII (resp. IX) we have proved the existence and unicity of the standard function Φ_w on G_w for any given type; the same follows now for G_∞, the standard function on G_∞ being $\otimes\Phi_w$. All the other results in theorem 4 of §49 (resp. in theorem 5 of §54) can now be extended trivially to G_∞. In particular, we will write:

$$I_\infty(g, \omega_\infty) = \int_{k_\infty^\times} \Phi(\begin{pmatrix} u & 0 \\ 0 & 1 \end{pmatrix}.g)\omega_\infty(u)d^\times u \ .$$

Moreover, for each infinite place w, we write G_w, e_w, for the "gamma factor" and the locally constant function on Ω_{k_w} which occur in the functional equation of Chapter VIII (resp. IX), and put

$$G_\infty(\omega_\infty) = \prod_\omega G_w(\omega_w),\ e_\infty(\omega_\infty) = \prod_\omega e_w(\omega_w) \ ,$$

$$J_\infty(g, \omega_\infty) = G_\infty(\omega_\infty)^{-1} I_\infty(g, \omega_\infty) \ ;$$

G_∞, e_∞ will be called the gamma factor and the locally constant factor for the given type. Then we have the functional equation

$$J_\infty(jg, \alpha_\infty^{-1}\omega_\infty^{-1}) = e_\infty(\omega_\infty) J_\infty(g, \omega_\infty) \ .$$

59. Let a divisor $\mathfrak{a}$ and a quasicharacter α of $k_A^\times/k^\times$ be given as in §§11-12; let an admissible type be given on G_∞, compatible with α_∞ in the obvious sense. By an automorphic form on G_A, of the type defined by these data, we understand a B-moderate function on G_A, satisfying conditions (A) to (D) of §12, and admissible of the given type on G_∞; here condition (E) of §14 need not be mentioned any more, since it is included in the requirements for an admissible function on G_∞.

Once for all, we select on G_∞ one standard function T of the given type, which we call the typical function (for that type). On B_∞, T induces a function of the form $(x, y) \longrightarrow \psi_\infty(y)W(x)$, where W is a V-valued function on $k_\infty^\times$ which we call the Whittaker function.

Using the results of §39, b) and c), or the corresponding results for the type $\mathcal{H}$ in Chapter IX, it is also easy to show that admissible functions on G_∞, left-invariant under $\begin{pmatrix} 1 & y \\ 0 & 1 \end{pmatrix}$ for all $y \in k_\infty$, lie in a finite-dimensional space; one can choose a basis for that space, consisting of functions $S_1, \ldots, S_m$, each of which induces on B_∞ a function of the form

$$(x, y) \longrightarrow C(x) \prod_w f_w(x_w) \ ,$$

where C is a V-valued and locally constant function on $k_\infty^\times$, and f_w is either a quasicharacter of $k_w^\times$ or the product of a quasicharacter of $k_w^\times$ and of $\log|x_w|$. We will write $U_i(x)$ for the functions thus induced on B_∞ by the functions S_i for $1 \leqq i \leqq m$.

60. A type on G_A being thus fixed once for all, let Φ be an automorphic form (of that type) on G_A; as before, we write F for the function induced by Φ on B_A; this can be expressed by the Fourier series (4) of Chapter III, with coefficients given by Fourier's formulas (6). Write now

$$(39) \qquad \Phi_o(g) = \int_{k_A/k} \Phi\left(\begin{pmatrix} 1 & y \\ 0 & 1 \end{pmatrix} \cdot g\right) dy, \quad \Phi_1(g) = \int_{k_A/k} \psi(-y)\Phi\left(\begin{pmatrix} 1 & y \\ 0 & 1 \end{pmatrix} \cdot g\right) dy .$$

Lemma 9. <u>On</u> G_∞, <u>for fixed values of the coordinates</u> g_v <u>of</u> g <u>at the finite places of</u> k, <u>the functions</u> Φ_o, Φ_1 <u>defined by</u> (39) <u>are admissible, and</u> Φ_1 <u>is standard.</u>

Clearly, all we need to show is that Φ_o, Φ_1 are B-moderate on G_w for each infinite place w, and that the integrals defining them may be differentiated inside the integral sign. Take a basis $(\eta_1, \ldots, \eta_n)$ of k over $\mathbf{Q}$; then $y \in k_A$ can be written as $\Sigma \eta_i u_i$ with $u_i \in \mathbf{Q}_A$ for $1 \leqq i \leqq n$. Write I for the interval $[0, 1[$ on $\mathbf{R}$, and $\overline{I}$ for its closure, i.e. $[0, 1]$; $I \times \prod \mathbf{Z}_p$ is a full set of representatives for $\mathbf{Q}_A/\mathbf{Q}$ in $\mathbf{Q}_A$, so that the integrals in (39) may be written as integrals over $(u_1, \ldots, u_n)$, each variable ranging over $I \times \prod \mathbf{Z}_p$. As Φ satisfies conditions (C), (D), it is locally constant as a function of the coordinates of the u_i at the finite places p of $\mathbf{Q}$ (cf. §12, Remark 1), so that the integration over the compact set $\prod \mathbf{Z}_p$ with respect to these coordinates amounts to a finite sum; each term in this sum is then an integral in the coordinates $u_i' = (u_i)_\infty$ of the u_i at the place ∞ of $\mathbf{Q}$, taken over I^n, or, what amounts to the same, over $\overline{I}^n$. As admissibility implies real analyticity (cf. §38), it is now clear that the integrals may be differentiated inside the integral sign. Now we have to show that Φ_o, Φ_1 are B-moderate on each G_w; writing the integrals which define them as we have just explained, we have to evaluate for

$p_w \longrightarrow +\infty$, a finite number of integrals of the form

$$\int |\Phi(\begin{pmatrix}1 & y_\infty\\0 & 1\end{pmatrix}\cdot\begin{pmatrix}p & 0\\0 & 1\end{pmatrix}\cdot g)|du'_1 \dots du'_n$$

with p given by $p = (p_v)$, $p_v = 1$ for $v \neq w$, $p_w > 0$; y_∞ is defined as usual, for $y = \Sigma\eta_i u_i \in k_A$. Here the integrand can be written

$$|\Phi(\begin{pmatrix}p & 0\\0 & 1\end{pmatrix}\cdot\begin{pmatrix}1 & p_\infty^{-1}y_\infty\\0 & 1\end{pmatrix}\cdot g)| \;;$$

it is understood that all the coordinates of g, except g_w, are being kept constant; g_w lies in a fixed compact subset of G_w, and so does y_∞ in k_∞ since $u'_i \in \overline{I}$ for $1 \leqq i \leqq n$. If we assume, as we may, $p_w \geqq 1$, $p_\infty^{-1}y_\infty$ also lies in a fixed compact subset of k_∞. Our conclusion follows now from lemma 8.

Lemma 9 shows in particular that $\Phi_1(g)$ is of the form $c_g T(g_\infty)$, where T is the typical function chosen above, and c_g depends solely upon the coordinates of g at the finite places of k; similarly, Φ_o must be a linear combination of the functions S_i defined in §59, with coefficients which depend solely upon these same coordinates. Now the Fourier formulas (6) of §13 may be written:

$$c_o(x) = \Phi_o(\begin{pmatrix}x & 0\\0 & 1\end{pmatrix}), \quad c(x) = \Phi_1(\begin{pmatrix}d^{-1}x & 0\\0 & 1\end{pmatrix}) \;.$$

As $d_\infty = 1$, and as $T((x_\infty, 0))$ is the Whittaker function $W(x_\infty)$, this shows that $c(x)$ may be written as $c(\mathrm{div}\, x)W(x_\infty)$, where $\mathfrak{m} \longrightarrow c(\mathfrak{m})$ is a mapping of $\mathfrak{M}$ into $\mathbb{C}$, and $c(\mathfrak{m}) = 0$ unless the divisor $\mathfrak{m}$ is positive. Similarly, c_o must be of the form $c_o(x) = \Sigma c_i(\mathrm{div}\, x)U_i(x_\infty)$, the U_i being as in §59.

61. We shall make essential use of the estimates given in the "convergence lemma" of Chapter V; these will be made more explicit now, for the case which is relevant here. For any $x \in k_A^\times$, we put

$$(40) \qquad \Sigma(x) = \sum_{\xi \in k^\times} |c(\operatorname{div} d\xi x) W(\xi x_\infty)| .$$

Lemma 10. _If Φ is B-moderate, we have $c(\mathfrak{m}) = O(|\mathfrak{m}|^{-a})$ for some $a \geqq 0$. If $c(\mathfrak{m}) = O(|\mathfrak{m}|^{-a})$ with $a \geqq 0$, the Fourier series (6) for F is absolutely convergent, uniformly over compact sets, and there are constants $\beta \geqq 0$, $\lambda > 0$, $C > 0$, $C' > 0$ such that_

$$\Sigma(x) \leqq C|x|^{-\beta} \text{ for } |x| \leqq 1 ,$$
$$\Sigma(x) \leqq C' \exp(-\lambda |x|^{1/n}) \text{ for } |x| \geqq 1 .$$

Moreover, similar estimates are valid if W is replaced by any one of its derivatives, and the series for F may be differentiated indefinitely term by term.

In fact, the "typical function" T is given by a tensor-product $\otimes \Phi_w$ of standard functions for the factors G_w of G_∞; to each, we can apply theorem 4(i) of §49 (resp. theorem 5(i) of §54). All conditions are now fulfilled for the application of lemma 4 of Chapter V (the "convergence lemma"), the conclusion being as stated above. We have put again $n = [k : \mathbb{Q}]$.

In particular, if Φ is any automorphic form and the coefficients $c(\mathfrak{m})$ are as defined above, the extended Dirichlet series

$$Z(\omega) = \Sigma c(\mathfrak{m}) \omega(\mathfrak{m})$$

is convergent for $\sigma(\omega)$ large; it will be called _the Dirichlet series attached to Φ_.

As in Chapters IV and VII, we also associate with Φ the function Φ' given by (7) of §17, i.e. by

$$\Phi'(g) = \Phi(g\mathfrak{d})\alpha(\det g)^{-1} ;$$

in view of Remark 3, §49, this is also an automorphic form, of a type determined by that of Φ; as typical function for that type, we select the function T' given by

$$T'(g) = T(g\mathfrak{a}_\infty)\alpha_\infty(\det g)^{-1} = T(g)M_\infty(\mathfrak{a}_\infty)\alpha_\infty(\det g)^{-1} ;$$

the corresponding Whittaker function is then

$$W'(x) = W(x)M_\infty(\mathfrak{a}_\infty)\alpha_\infty(x)^{-1} .$$

We will call (Φ, Φ') an <u>automorphic pair</u>, and write $Z'(\omega)$ for the Dirichlet series attached to Φ'.

62. We write k_∞^1 for the subgroup $k_\infty^\times \cap k_A^1$ of $k_\infty^\times$, i.e. for the kernel of the restriction of $u \longrightarrow |u|_A$ to $k_\infty^\times$. On the other hand, for each $\nu \in \mathbf{R}_+^\times$, define u_ν as the element (u_w) of $k_\infty^\times$ given by $u_w = \nu$ for each infinite place w; call N the image of $\mathbf{R}_+^\times$ under $\nu \longrightarrow u_\nu$. Then we have $|u_\nu| = \nu^n$, $k_\infty^\times = k_\infty^1 \times N$ and $k_A^\times = k_A^1 \times N$.

On $k_v^\times$, for any finite v, we have normalized the Haar measure so that the measure of $r_v^\times$ is 1; on $k_w^\times$, for any infinite w, we have normalized it (in §46) by $d^\times u_w = |u_w|_w^{-1} du_w$, where the additive Haar measure du_w has been normalized as explained there. This defines normalized Haar measures on $k_\infty^\times$ and on $k_A^\times$. On N, we take the Haar measure $d^\times \nu = \nu^{-1} d\nu$; on k_∞^1 and k_A^1, we normalize the Haar measures $d_1 u$ so that $d^\times u$ (on $k_\infty^\times$ resp. $k_A^\times$) is the product of $d_1 u$ and $d^\times \nu$.

Each quasicharacter of $k_A^\times/k^\times$ may be regarded as given by the character it induces on the compact group $k_A^1/k^\times$ and by the quasicharacter it induces on N; the latter is of the form $\nu \longrightarrow \nu^z$ with $z \in \mathbf{C}$. In particular, it is of the form ω_s if and only if it is trivial on $k_A^1/k^\times$, and then it induces $\nu \longrightarrow \nu^{ns}$ on N. This may also be expressed by saying that Ω_k is the direct product of the (discrete) group of characters of $k_A^1/k^\times$ and of the group of quasicharacters of N, the latter being isomorphic to $\mathbf{C}$. For any $\omega \in \Omega_k$, $|\omega|$ may be written as ω_σ with $\sigma \in \mathbf{R}$, and then we write $\sigma = \sigma(\omega)$ (cf. §46).

From now on, we will usually write $\mathfrak{f}_o$ for the conductor of the quasicharacter $\omega \in \Omega_k$; this depends only upon the connected component of Ω_k to which ω belongs; for each such component, we choose an idele $f_o = (f_v)$ (the same as the one denoted by f in §8) such that $f_v = 1$ whenever v is not a place in $\mathfrak{f}_o$ (and in particular at all infinite places) and that $\mathfrak{f}_o = \mathrm{div}(f_o)$. We will write $\kappa_v(\omega)$ for the "normalized Gaussian sum" κ_v attached to κ_v at v (cf. §10); this is 1 at all finite places not occurring in $\mathfrak{f}_o$; at an infinite place w, it is the same as the number $\kappa(\varphi_w)$ occurring in Tate's lemma (Chapter VIII, §46) and in §§47-49. We write $\kappa(\omega)$ for the product $\prod \kappa_v(\omega)$ taken over all the places of k, and $\kappa_o(\omega)$, $\kappa_\infty(\omega)$ for the partial products taken over the finite and over the infinite places, respectively. We write again $\varepsilon(\omega)$ for the constant factor in the functional equation for the L-function attached to ω; this is given by $\varepsilon(\omega) = \kappa(\omega)\omega(df_o)$.

63. We are now (at last) ready for our main task, which is to prove the functional equation for the extended Dirichlet series attached to an automorphic pair, and conversely to give the conditions for two such series to be those attached to such a pair. We consider the integral (similar to (14) of Chapter VII, §26):

$$(41) \qquad I(f, e, t, \omega) = \int_{k_A^1} c(\mathrm{div}\, dftu) W(f_\infty t_\infty u_\infty) \psi(etu) \omega(u) d_1 u \ .$$

Here, as in §26, (f, e) is assumed to be reduced in the sense of §16. With $\Sigma(x)$ defined by (40), this is majorized by

$$\int_{k_A^1} |c(\mathrm{div}\, dftu) W(f_\infty t_\infty u_\infty)| d_1 u = \int_{k_A^1/k^\times} \Sigma(ftu) d_1 u \ .$$

As $k_A^1/k^\times$ is compact, lemma 10 shows that this is convergent and gives a bound on its order of magnitude; therefore (41) is absolutely convergent and is $O(|t|^{-\beta})$ for $|t| \longrightarrow 0$ and $O(\exp(-\lambda|t|^{1/n}))$ for $|t| \longrightarrow +\infty$, and we have:

$$(42)\qquad \begin{aligned} I(f, e, t, \omega) &= \int_{k_A^1/k^\times} [\sum_{\xi\in k^\times} c(\operatorname{div} dft\xi u)W(f_\infty t_\infty \xi u_\infty)\psi(et\xi u)]\omega(u)d_1u \\ &= \int_{k_A^1/k^\times} [F(ftu, etu) - c_0(ftu)]\omega(u)d_1u . \end{aligned}$$

Here, too, we observe that $I(f, e, t, \omega)$ is 0 unless the conductor $\mathfrak{f}_0$ of ω divides $\mathfrak{f} = \operatorname{div}(f)$; this is similar to the first part of proposition 7, §26, and is proved in the same way. In fact, take a place v in $\mathfrak{f}$, and replace u by $u(1 + f_v s)$, with $s \in r_v$, in (41) or (42); this does not change the integral; on the other hand, it affects only the factor $\omega(u)$ in the integrand, multiplying it with $\omega(1 + f_v s)$; therefore the latter factor must be 1 for all $s \in r_v$, as was to be proved.

Now consider the integral

$$(43)\qquad J(\omega) = \int_0^{+\infty} I(f, e, tu_\nu, \omega)\omega(tu_\nu)d^\times\nu \ ;$$

the above estimate for $I(f, e, t, \omega)$ shows that this is absolutely convergent if it is so near $\nu = 0$, and that this is the case if $\sigma(\omega)$ is large enough. Clearly we can also write

$$(44)\qquad J(\omega) = \int_{k_A^\times} c(\operatorname{div} dft)W(f_\infty t_\infty)\psi(et)\omega(t)d^\times t$$

provided the latter integral is absolutely convergent; this can be seen, just as above, by observing that it is majorized by

$$\int_{k_A^\times/k^\times} \Sigma(ft)\,|\omega(t)|\,d^\times t = \int_0^{+\infty}\left[\int_{k_A^1/k^\times} \Sigma(ftu_\nu)d^\times t\right].\ |\omega(u_\nu)|\,d^\times\nu \ ,$$

and applying lemma 10, which shows that it is convergent for $\sigma(\omega)$ large enough.

Proposition 14. <u>The integral</u> $J(\omega)$ <u>given by</u> (42) <u>is</u> 0 <u>unless the conductor</u> $\mathfrak{f}_0$ <u>of</u> ω <u>divides</u> $\mathfrak{f} = \mathrm{div}(f)$; <u>if</u> $\mathfrak{f}_0 = \mathfrak{f}$, <u>and if</u> $\sigma(\omega)$ <u>is large enough,</u> $J(\omega)$ <u>is given by</u>

$$J(\omega) = \kappa_0^{-1}\omega(df_0)^{-1}|\mathfrak{f}|^{-1/2}\left[\prod_{v/\mathfrak{f}}(1 - q_v^{-1})\omega_v(-e_v)\right]^{-1}.$$

$$[\Sigma c(\mathfrak{m})\omega(\mathfrak{m})]\int_{k_\infty^\times} W(uf_\infty)\psi_\infty(ue_\infty)\omega_\infty(u)d^\times u \quad .$$

The first assertion follows from the corresponding one about $I(f, e, t, \omega)$. As to the latter, we only have to observe that the integral (44) splits into local factors for the finite places v of k, and an integral taken over $k_\infty^\times$; the former are precisely the same as in the function-field case (cf. the proof of proposition 7, §26). The result is the one given above. In view of this, the absolute convergence of $J(\omega)$ implies that of the series $Z(\omega)$, and of the integral on $k_\infty^\times$. The latter is clearly a special case of the one occurring in the functional equation for standard functions (cf. §58), and can be written as:

$$I_\infty(b_\infty, \omega_\infty) = \int_{k_\infty^\times} T\left(\begin{pmatrix} u & 0 \\ 0 & 1\end{pmatrix}\right).b_\infty)\omega_\infty(u)d^\times u \quad ,$$

with $b_\infty = (f_\infty, e_\infty)$.

64. As in Chapter VII, §27, we say that the automorphic form Φ is B-<u>cuspidal</u> if it satisfies the condition

$$\int_{k_A/k} \Phi\left(\begin{pmatrix} x & y \\ 0 & 1\end{pmatrix}\right)dy = 0$$

for all x, i.e. if the constant term $c_0(x)$ in the Fourier series for F

is 0. We can now prove the analogue of theorem 2 of §27:

Theorem 6. Let Φ, Φ' be a B-cuspidal automorphic pair on G_A; let $Z(\omega)$, $Z'(\omega)$ be the extended Dirichlet series respectively attached to Φ and to Φ'. Let $G_\infty(\omega_\infty)$, $e_\infty(\omega_\infty)$ be the gamma factor and the locally constant factor attached to the type of Φ. Then, on the group of the quasicharacters of $k_A^\times/k^\times$ whose conductor $\mathfrak{f}_0$ is disjoint from $\mathfrak{a}$, we have the functional equation

$$G_\infty(\omega_\infty)Z(\omega) = \omega_\infty(-1)e_\infty(\omega_\infty)a(\mathfrak{f}_0)\omega(\mathfrak{a})\varepsilon(\omega)^2 G_\infty(a_\infty^{-1}\omega_\infty^{-1})Z'(\omega^{-1}) ,$$

in the sense that both sides can be continued as entire functions to the whole of that group and are equal there; moreover, on each connected component of that group, their common value is bounded within each strip $A \leqq \sigma(\omega) \leqq B$.

In Chapter IV, it has been proved that, if Φ, Φ' are an automorphic pair, the functions F, F' they induce on B_A must fulfill condition (II') of §17, with b, b', $\mathfrak{b}$, $\mathfrak{z}$ as in proposition 4 of §17. A simple calculation shows that this can be written as follows:

$$(45)\qquad F(tf, te) = F'(at^{-1}f, at^{-1}e')M'_\infty(\mathfrak{b}_\infty)a_\infty(f_\infty)a(\mathfrak{f})$$

with $\mathfrak{f} = \mathrm{div}(f)$. Now restrict ω to a given connected component $\mathfrak{C}$ of Ω_k, with a conductor $\mathfrak{f}_0$ disjoint from $\mathfrak{a}$. Then choose f so that $\mathfrak{f} = \mathfrak{f}_0$, replace t by tu in the above formula, multiply both sides with $\omega(u)$, and integrate over $k_A^1/k^\times$. As $c_0(x)$ is assumed to be 0, (42) shows that we get $I(f, e, t, \omega)$ in the left-hand side. A similar observation applies to the right-hand side; therefore, writing I' for the integral derived from F' just as I is derived from F, we get

$$(46)\qquad I(f, e, t, \omega) = I'(f, e', at^{-1}, \omega^{-1})M'_\infty(\mathfrak{b}_\infty)a_\infty(f_\infty)a(\mathfrak{f}_0) .$$

The left-hand side is $O(\exp(-\lambda|t|^{1/n}))$ for $|t| \longrightarrow +\infty$; the right-hand side is $O(\exp(-\lambda|t|^{-1/n}))$ for $|t| \longrightarrow 0$. Consequently, if now we replace t by tu_ν, multiply with $\omega(tu_\nu)$ and integrate with the measure $d^\times\nu$ over $R_+^\times$, we get an integral which is always absolutely convergent and defines an entire function of ω on $\mathfrak{G}$. As the left-hand side shows, this integral is no other than $J(\omega)$, which therefore, under the present circumstances, defines an entire function on $\mathfrak{G}$; in view of proposition 14, we may say that this gives the analytic continuation of $Z(\omega)I_\infty(b_\infty, \omega_\infty)$ to $\mathfrak{G}$. On the other hand, if we treat in the same manner the right-hand side of (46), we get an integral $J'(\omega^{-1})$ similar to $J(\omega)$. Of course this is also an entire function on $\mathfrak{G}$. At the same time, for $\sigma(\omega^{-1})$ large enough, it can be expressed in terms of $Z'(\omega^{-1})I'_\infty(b'_\infty, \omega^{-1})$, with $b'_\infty = (f_\infty, e'_\infty)$, again by means of proposition 14; by I'_∞, we have denoted the integral, similar to I_∞, derived from the typical function T' for the type of Φ on G_∞ (cf. §61). Putting these results together, and taking into account the relations between f, e, e', a in proposition 4 of §17, and the definition of f_o in §62, we get in the first place:

$$(47)\qquad Z(\omega)I_\infty(b_\infty, \omega_\infty) = \gamma Z'(\omega^{-1})I'_\infty(b'_\infty, \omega^{-1})M'_\infty(\not{b}_\infty)\alpha_\infty(f_\infty)$$

with γ given by

$$\gamma = \kappa_o^2\omega(df_o)^2\omega(\mathfrak{d})\alpha(\not{f}_o) .$$

At the same time, since $a_w = 1$ at all infinite places, the relations in proposition 4 of §17 give $b_\infty = -jb'_\infty\not{b}_\infty\mathfrak{d}_\infty$, and therefore, in view of the definition of T' in §61, and of the fact that it belongs to the representation M'_∞ of $\mathfrak{K}_\infty$:

$$I_\infty(jb_\infty, \alpha_\infty^{-1}\omega_\infty^{-1}) = \int T(\begin{pmatrix} u & 0 \\ 0 & 1\end{pmatrix}.b'_\infty\not{b}_\infty\mathfrak{d}_\infty)(\alpha_\infty^{-1}\omega_\infty^{-1})(u)d^\times u$$

$$= \alpha_\infty(f_\infty)\int T'(\begin{pmatrix} u & 0 \\ 0 & 1\end{pmatrix}.b'_\infty\not{b}_\infty)\omega_\infty(u)^{-1}d^\times u = I'_\infty(b'_\infty, \omega_\infty^{-1})M'_\infty(\not{b}_\infty)\alpha_\infty(f_\infty) ,$$

the integrals being taken over $k_\infty^\times$. Thus the above formula can be rewritten as

$$Z(\omega)I_\infty(b_\infty, \omega_\infty) = \gamma Z'(\omega^{-1})I_\infty(jb_\infty, \alpha_\infty^{-1}\omega^{-1}) .$$

If in this formula $b_\infty = (f_\infty, e_\infty)$ is replaced by (tf_∞, te_∞) with any $t \in k_\infty^\times$, both sides are multiplied with $\omega_\infty(t)^{-1}$, as follows at once from (41) and the fact that the typical function T belongs to the quasicharacter α_∞ of $\mathfrak{Z}_\infty$; for a similar reason, both sides are multiplied with the same factor if b_∞ is replaced by $b_\infty \mathfrak{p}'_\infty \mathfrak{z}_\infty$ with any $\mathfrak{p}'_\infty \mathfrak{z}_\infty$ in $\mathfrak{K}_\infty \mathfrak{Z}_\infty$. Since we were allowed to choose for (f_w, e_w), for each infinite place w, any "reduced" pair in the sense of §16, this shows that we have, for any $g \in G_\infty$:

$$Z(\omega)I_\infty(g, \omega_\infty) = \gamma Z'(\omega^{-1})I_\infty(jg, \alpha^{-1}\omega^{-1}) ,$$

both sides still being entire functions on $\mathfrak{G}$. With the notations of §58, this may also be written as

$$G_\infty(\omega_\infty)Z(\omega)J_\infty(g, \omega_\infty) = \gamma G_\infty(\alpha_\infty^{-1}\omega_\infty^{-1})Z'(\omega^{-1})J_\infty(jg, \alpha_\infty^{-1}\omega_\infty^{-1}) .$$

It follows from theorem 4(iv) of Chapter VIII (resp. theorem 5(iv) of Chapter IX) that, for each ω_∞, we can choose g so that $J_\infty(g, \omega_\infty)$ is not 0. Consequently $G_\infty(\omega_\infty)Z(\omega)$ is everywhere holomorphic in ω, and the functional equation for J_∞, as it has been stated in §58, gives at once the functional equation for Z in theorem 6. Finally, theorem 4(iv) (resp. 5(iv)) shows also that one can choose a distribution T with the support $\{1_2\}$ on G_∞, so that $T[J_\infty(g, \omega_\infty)]$ is a non-zero polynomial function on $\mathfrak{G}$. As we have shown, $Z(\omega)I_\infty(b_\infty, \omega_\infty)$ and consequently also $Z(\omega)I_\infty(g, \omega_\infty)$ can be obtained (up to constant factors) by integrating over N the function $I(f, e, tu_\nu, \omega)\omega(tu_\nu)$, which decreases exponentially for $\nu \longrightarrow 0$ and $\nu \longrightarrow +\infty$. Combining now lemma 10 of §61 with routine arguments, such as have been repeatedly used above, one finds that $T[Z(\omega)I_\infty(g, \omega_\infty)]$ may be calculated by differentiating formally the integrals in question and that similar estimates remain valid for the differentiated formulas; therefore $T[Z(\omega)I_\infty(g, \omega_\infty)]$ remains bounded within each strip $A \leqq \sigma(\omega) \leqq B$ in $\mathcal{C}$. This is the same as $G_\infty(\omega_\infty)Z(\omega)T[J_\infty(g, \omega_\infty)]$, and the last factor in the latter product is a non-zero polynomial function, hence bounded from below in absolute value when ω tends to infinity in $\mathcal{C}$. This completes the proof of the last assertion in our theorem.

65. As in Chapter VII, we can treat similarly the case when Φ, Φ' are not assumed to be B-cuspidal, by taking into account the term $-c_o(ftu)$ in the last integral in (42) and the corresponding term from Φ'. We will skip most of the calculations and merely describe the main steps and the final result, which is similar to proposition 8 of Chapter VII, §28.

As we have seen, $c_o(x)$ can be written as $\Sigma c_i(\mathrm{div}\, x)U_i(x_\infty)$, where the U_i are as in §59. Combining this with the fact that $c_o(x)$ is constant on the cosets of $k^\times \prod r_v^\times$ in $k_A^\times$, one concludes easily that it is a linear combination, with constant scalar coefficients, of the functions of the form $C\theta(x)$ or $C\theta(x) \log|x|$, where: (a) θ is a quasi-character of $k_A^\times / k^\times \prod r_v^\times$ (i.e. an unramified quasicharacter of $k_A^\times/k^\times$); (b) C is a constant vector in V; (c) $C\theta_\infty(x_\infty)$ resp. $C\theta_\infty(x_\infty) \log|x_\infty|$ is one of the functions U_i; moreover, if $C\theta(x) \log|x|$ is such, so is $C\theta(x)$; we will denote by $C_j\theta_j$ those of the form $C\theta$.

Of course the same holds true for the type of Φ', so that we may write the constant term $c_o'(x)$ in the Fourier series for F' as a linear combination of finitely many functions $C_h'\theta_h'(x)$ resp. $C_h'\theta_h'(x) \log|x|$.

As in the proof of theorem 6, let us restrict ω to one component

$\mathfrak{C}$ of Ω_k, with the conductor $\mathfrak{f}_0$ disjoint from $\mathfrak{A}$; we apply (45), with f chosen so that div f $= \mathfrak{f}_0$, and combine this with (42) and the similar formula for I' and F'. In (42), the part of the integral arising from the term $-c_0(ftu)$ is 0 unless ω coincides with some θ_j^{-1} on $k_A^1/k^\times$, i.e. unless some θ_j^{-1} lies in $\mathfrak{C}$, in which case $\mathfrak{f}_0$ is 1. Similarly, when we apply (42) to $I'(f, e', at^{-1}, \omega^{-1})$, the term arising from c'_0 is 0 unless some θ'_h lies in $\mathfrak{C}$, which again implies $\mathfrak{f}_0 = 1$. Consequently, if $\mathfrak{C}$ contains none of the θ_j^{-1}, θ'_h, and in particular if $\mathfrak{f}_0 \neq 1$, we can proceed as in the proof of theorem 6, the conclusions being the same as before.

Now assume that $\mathfrak{C}$ contains at least one θ_j^{-1} or one θ'_h; then $\mathfrak{f}_0 = 1$; we may take $f_v = 1$, $e_v = e'_v = 0$ at all finite places, and κ_0 is 1. Combining (45) with (42) and the similar formula for I', F', we get a formula similar to (46), but with added terms arising from c_0, c'_0. Replace t by tu_ν, multiply with $\omega(tu_\nu)$, and integrate with the measure $d^\times\nu$ from 1 to $+\infty$; the integral arising from I is then an entire function of ω on $\mathfrak{C}$. As to the other terms, they are absolutely convergent for $-\sigma(\omega)$ large enough, and then one can evaluate explicitly those arising from c_0, c'_0; if for instance c_0 contains a term $C\theta(x)$ resp. $C\theta(x) \log|x|$, and θ^{-1} is in $\mathfrak{C}$, we may put $\omega = \theta^{-1}\omega_s$, and the part of the integral coming from that term is of the form $C|t|^s\lambda s^{-1}$ resp. $C|t|^s(\lambda s^{-1} + \mu s^{-2})$, with scalar constants λ, μ. On the other hand, if we integrate from 0 to 1 with the measure $d^\times\nu$, the integral arising from I' is an entire function; the other terms are absolutely convergent for $\sigma(\omega)$ large, and those arising from c_0, c'_0 are formally the negatives of the earlier ones, e.g. $-C|t|^s\lambda s^{-1}$ for the part arising from $C\theta(x)$. Combining those results with proposition 14, one finds in the first place that $G_\infty(\omega_\infty)Z(\omega)$ and $G_\infty(\alpha_\infty^{-1}\omega_\infty^{-1})Z'(\omega^{-1})$ can be continued to $\mathfrak{C}$ as meromorphic functions and satisfy the same functional equation as before, i.e. the one in theorem 6. One also finds that they have a double pole at θ^{-1} whenever c_o contains a term $C\theta(x) \log|x|$, a simple one if c_o contains no such term but a term $C\theta(x)$, and similarly a double or simple pole or none at θ' according as c'_o contains a term $C'\theta'(x) \log|x|$, a term $C'\theta'(x)$, or none. Finally, by making use of a distribution T with the support $\{1_2\}$ on G_∞ just as in the proof of theorem 6, one shows that the meromorphic functions in question are bounded in each strip $A \leqq \sigma(\omega) \leqq B$ in $\mathcal{C}$ outside circles around the poles.

Remark 1. In general, for a given type, there is no function $C\theta$ fulfilling the conditions (a), (b), (c). More precisely, when α is given, there are only enumerably many types for which there are such functions.

Remark 2. As a detailed calculation would show, the residues of $G_\infty(\omega_\infty)Z(\omega)$, and therefore the coefficients $c(\mathfrak{m})$, determine c_o, c'_o uniquely. One finds also that a term $C\theta$ cannot occur in c_o unless $J_\infty(g, \theta_\infty^{-1})$ is (up to a constant factor) the same as one of the functions S_i. This excludes some of the functions $C\theta$ which fulfill conditions (a), (b), (c); for instance, if for some place w where $k_w = \mathbb{C}$ the type is $(1, 0, 0, 2)$, this rule excludes the same functions which would be excluded if that type were replaced by $\mathcal{H}_{\mathbb{C}}$, as explained in §52.

Remark 3. Had we replaced the condition $W^{h+1}\varphi_1 = 0$ by $W^n\varphi_1 = 0$ in the definition of admissible functions of discrete type in §41, c), this (as observed there, Remark 1) would not have affected the concept of a standard function, but it would allow for more functions S_i, U_i. With this modified definition, whenever $C\theta$ is one of the functions $C_j\theta_j$, i.e. whenever it has the properties (a), (b), (c), $C\omega_1\theta^{-1}$ is one of the functions $C'_h\theta'_h$. However, this advantage is nullified by the fact that the additional functions $C\theta$ allowed by the new definition are all excluded by the rule mentioned in Remark 2.

66. The following lemmas will be needed for dealing with the converse of theorem 6.

Lemma 11. For $z = x + iy$ and any interval $I = [a, b]$ in R, let S be the half-strip $x \in I$, $y \geqq 0$. Let f be a holomorphic function in S; assume that $|f| \leqq C$ on the boundary of S and $|f| \leqq C'e^{\gamma y}$ in S, with constants C, C', $\gamma > 0$. Then $|f| \leqq C$ in S.

This is a well-known consequence of the Phragmèn-Lindelöf principle. Choose $\lambda > 0$ so that $|2\lambda a| < \pi$ and $|2\lambda b| < \pi$; then, for any $\varepsilon > 0$, consider the function $|f(z)\exp(-\varepsilon e^{-i\lambda z})|$; this is $\leqq C$ on the boundary of S and tends to 0 for $y \longrightarrow +\infty$, uniformly in x; therefore it is $\leqq C$. As this is so for all $\varepsilon > 0$, the conclusion follows.

Lemma 12. Let φ, φ' be two continuous functions on $R_+^\times$, such that both $\varphi(\nu)$ and $\varphi'(\nu^{-1})$ are $O(e^{-A\nu})$, with some $A > 0$, for $\nu \longrightarrow +\infty$, and $O(\nu^{-B})$, with some $B \geqq 0$, for $\nu \longrightarrow 0$. Put

$$f(s) = \int_0^{+\infty} \varphi(\nu)\nu^s d^\times\nu\ , \qquad f'(s) = \int_0^{+\infty} \varphi'(\nu)\nu^s d^\times\nu\ ;$$

then these integrals are absolutely convergent and define holomorphic functions f, f' in the half-planes $\mathrm{Re}(s) > B$ and $\mathrm{Re}(s) < -B$, respectively. Assume now that, for some $\sigma > B$ and some $\sigma' < -B$, $t \longrightarrow f(\sigma + it)$ and $t \longrightarrow f'(\sigma' + it)$ are $O(|t|^{-2})$. Then we have $\varphi = \varphi'$ if and only if f, f' can be continued to one and the same entire function in the s-plane, bounded in every strip $\sigma_1 \leqq \mathrm{Re}(s) \leqq \sigma_2$.

The first part is obvious; it is also obvious, if $\varphi = \varphi'$, that the integrals for f and f' are the same, that they are everywhere absolutely convergent, and that they define an entire function, bounded in every strip. As to the converse, replace s by $\sigma + 2\pi it$ and ν by e^x in the integral for f; then we see that $t \longrightarrow f(\sigma + 2\pi it)$ is the Fourier transform of $x \longrightarrow \varphi(e^x)e^{\sigma x}$. In view of the assumption on the order of magnitude of $f(\sigma + 2\pi it)$, this implies that we have

$$\varphi(e^x)e^{\sigma x} = \int_{-\infty}^{+\infty} f(\sigma + 2\pi i t)e^{-2\pi i t x}dt \quad ,$$

which may also be written as

(48)
$$\varphi(\nu) = \frac{1}{2\pi i}\int_{(\sigma)} f(s)\nu^{-s}ds \quad ,$$

where the integral is taken from $\sigma - i\infty$ to $\sigma + i\infty$ on the line $\mathrm{Re}(s) = \sigma$. Quite similarly, we get

$$\varphi'(\nu) = \frac{1}{2\pi i}\int_{(\sigma')} f'(s)\nu^{-s}ds \quad .$$

Now assume that f, f' can be continued to one and the same entire function F, bounded in $\sigma' \leqq \mathrm{Re}(s) \leqq \sigma$. In view of our assumptions on f, f', lemma 11 can be applied (with any $\gamma > 0$) to the function $F(s)s^2$ in the half-strip $\sigma' \leqq \mathrm{Re}(s) \leqq \sigma$, $\mathrm{Im}(s) \geqq 0$; therefore it is bounded there; similarly it is bounded in the lower half-strip $\sigma' \leqq \mathrm{Re}(s) \leqq \sigma$, $\mathrm{Im}(s) \leqq 0$. That being so, one can shift the line of integration from (σ) to (σ') in (48). This gives $\varphi = \varphi'$.

Lemma 12 is essentially due to Hecke. As he had observed, it can be extended in an obvious manner to the case when f, f' can be continued to one and the same meromorphic function with finitely many poles, bounded in every strip outside circles around the poles. Then, in shifting the line of integration from (σ) to (σ'), one has to take the residues of $f(s)\nu^{-s}$ into account; instead of being 0, $\varphi - \varphi'$ is a linear combination of the functions ν^{-a}, where a ranges over the poles of f, if these poles are simple; otherwise one has terms of the form $\nu^{-a}(\log \nu)^i$, with $0 \leqq i < m$ if a is a pole of order m. This would be needed if we were to prove the converse of the results of §65; since we will not do this, lemma 12 will suffice for our purposes.

67. In applying lemma 12, we shall also need some estimates for the functions $I_\infty(g, \omega_\infty)$, $J_\infty(g, \omega_\infty)$ which occurred in the proof of

theorem 6. These are given in the following lemma:

Lemma 13. Let χ be a fixed quasicharacter of $k_\infty^\times$. Then, for every $M \geqq 1$, every large σ and every $g \in G_\infty$, there is C such that $|I_\infty(g, \chi\omega_{\sigma+it})| \leqq C|t|^{-M}$; for every interval $[a, b]$ in $\mathbf{R}$ and every $g \in G_\infty$, there is C' and γ such that $|J_\infty(g, \chi\omega_{\sigma+it})| \leqq C' e^{\gamma|t|}$ for $a \leqq \sigma \leqq b$.

It is clearly enough to deal with one infinite place at a time, or in other words to prove our assertions for the functions $I(g, \omega)$, $J(g, \omega)$ of theorem 4, §49; moreover (just as in §49) if our assertions are valid for a "reduced type" (in the sense of Chapter VIII, §43), they remain so in general. Now choose a Schwartz function S in K^2 as in the proof of proposition 13, §48; with the notations of that proposition and of proposition 12, §47, $J(g, \omega)$ is the same as $P_\mathfrak{a}(\omega)^{-1} J(S, g, \omega)$ if the type is a discrete one, as $J(S, g, \omega)$ otherwise, and $I(g, \omega)$ is the same as $I(S, g, \omega)$; since $P_\mathfrak{a}(\omega)$ is a polynomial function, it will be enough if we prove our assertions for $I(S, g, \omega)$ and $J(S, g, \omega)$. These can be written as

$$I(\omega) = \int F(x, y)\omega(x)(\beta\omega)(y)d^\times x d^\times y \quad ,$$

$$J(\omega) = (\Delta_\omega \otimes \Delta_{\beta\omega})(F) \quad ,$$

where we have written β instead of $\omega_1 \mathfrak{a}$, and where F is a Schwartz function on K^2. We begin with $J(\omega)$, e.g. in the case $K = \mathbf{R}$. Take an integer $N \geqq 0$ such that $N + \sigma(\chi) + a > 0$, $N + \sigma(\beta\chi) + a > 0$. It is easily seen (e.g. by induction on N) that F can be written as a finite linear combination, with constant coefficients, of functions $e^{-\pi(x^2+y^2)}x^m y^n$, $e^{-\pi x^2}x^m G(y)$, $e^{-\pi y^2}y^n H(x)$ and of a function $F_1(x, y)$, where the functions G and H are Schwartz functions on $\mathbf{R}$ and the exponents of x and y, in the formal expansions of the functions G, H and F_1 at 0, are all $\geqq N$. In §46, we have seen that e.g. $\Delta_\omega(e^{-\pi x^2}x^m)$ is a polynomial function of ω, i.e. here a polynomial in $s = \sigma + it$; therefore $J(\omega)$ is a linear combination, with coefficients which are polynomials in s, of the functions 1, $\Delta_\omega(H)$, $\Delta_{\beta\omega}(G)$ and $(\Delta_\omega \otimes \Delta_{\beta\omega})(F_1)$. Now part (ii) of lemma 6, §46, shows that each one of the latter functions can be written as the product of a gamma factor and of an integral which is uniformly absolutely convergent, and therefore bounded, within the given range $a \leq \sigma \leq b$. As to the gamma factors $\mathcal{G}(\omega)^{-1}$, $\mathcal{G}(\beta\omega)^{-1}$ and $\mathcal{G}(\omega)^{-1}\mathcal{G}(\beta\omega)^{-1}$, the elementary properties of the gamma function show that they are uniformly $O(e^{\gamma|t|})$, for a suitable γ, within the given range. This proves our assertion concerning $J(\omega)$ in the case $K = \mathbf{R}$; the case $K = \mathbf{C}$ can be treated quite similarly, by replacing the functions $e^{-\pi x^2}x^m$ by the functions $e^{-2\pi x\bar{x}}x^m\bar{x}^n$, etc.

As to $I(\omega)$, we may clearly assume that $\sigma(\chi) > 0$, $\sigma(\beta\chi) > 0$, and consider the integral

$$I(t, t') = \int F(x, y)\chi(x)(\beta\chi)(y)\,|x|^{2\pi i t}|y|^{2\pi i t'}d^\times x\, d^\times y \ ,$$

which is absolutely convergent. Writing $K^\times$ as the product of $\mathbf{R}_+^\times$ and of the group $\{\pm 1\}$ if $K = \mathbf{R}$, the group $\{x \mid x\bar{x} = 1\}$ if $K = \mathbf{C}$, integrating over the compact factor of $(K^\times)^2$, and writing $|x| = e^u$, $|y| = e^{u'}$, we get I as a Fourier transform

$$I(t, t') = \int \Phi(u, u')e^{2\pi i(tu+t'u')}du\,du' \ .$$

Then the functions $t^p t'^q I(t, t')$ are the Fourier transforms of the partial derivatives of Φ; as F is a Schwartz function, they are all given by absolutely convergent integrals and are therefore bounded. In particular, $t^M I(t, t)$ is bounded for all M, as was to be proved.

Actually, a better estimate for $I(S, g, \omega)$ can be obtained by taking for S the particular function used in the proof of proposition 13, §48 (i.e. the Fourier transform of the function S^* defined there, with $E(t) = e^{-\pi t}$ resp. $e^{-2\pi t}$); one finds that, assumptions being as before, $I(S, g, \omega)$ is then $O(e^{-\delta|t|})$, with a suitable $\delta > 0$ (depending upon g). The proof of this depends upon the fact that the function F in the above formulas, i.e. essentially the function S'_g in the notation of §§44 and 47, can be continued as an entire function into the complexification of the vector-space K^2 over $\mathbf{R}$, so that one can push the integral into that complexification. One can also express this by saying that, in the above expression of $I(t, t')$ as a Fourier integral, one may, after continuing the function $\Phi(u, u')$ as a complex-analytic function into the complex (u, u')-space, replace u, u' by $u + i\xi$, $u' + i\xi'$, where ξ, ξ' are given sufficiently small constant values. Essentially trivial estimates for Φ show that this is permissible, and give the announced result.

68. In Chapter VII, we refrained from proving the converse of proposition 8, §28, and merely gave (in theorem 3) the converse of theorem 2, dealing with the case of a B-cuspidal automorphic pair. We will do the same here, and give the converse of theorem 6, omitting its generalization to other than B-cuspidal pairs; the latter, which presents no substantial difficulty, will be left to the reader.

As in Chapters VI and VII, we attach a Hecke operator to every divisor $\mathfrak{n}$ disjoint from $\mathfrak{a}$, and write $\Phi_{\mathfrak{n}}$, $Z_{\mathfrak{n}}$, etc., for the transforms of Φ, Z, etc. under this operator. Here, too, it is obvious that theorem 6 can be applied to $\Phi_{\mathfrak{n}}$ and $\Phi'_{\mathfrak{n}}$, so that the corresponding Dirichlet series $Z_{\mathfrak{n}}$, $Z'_{\mathfrak{n}}$ have the properties stated in that theorem; one merely has to replace the factor $\alpha(\mathfrak{f}_0)$ by $\alpha(\mathfrak{f}_0\mathfrak{n})$ in the functional equation.

A type of automorphic forms being given on G_A, and notations being as before, we attach to each extended Dirichlet series $Z(\omega) = \sum c(\mathfrak{m})\omega(\mathfrak{m})$ the function F on B_A defined by the Fourier series

(49) $$F(x, y) = \sum_{\xi \in k^\times} c(\operatorname{div} \xi dx) W(\xi x_\infty) \psi(\xi y) \;.$$

At the same time, another Dirichlet series Z' being given with the coefficients c', we attach to it the function F' defined by the Fourier series similar to the above one, with c replaced by c' and W replaced by the function W' defined in §61. Our next (and last) theorem gives the conditions that must be fulfilled by Z, Z' in order that F, F' may extend to an automorphic pair Φ, Φ'.

Theorem 7. Let S be a set of finite places of k, including those in $\mathfrak{a}$, with the approximation property. Let Ω_S be the group of the quasicharacters of $k_A^\times/k^\times$ whose conductor is disjoint from S. Let $Z(\omega)$, $Z'(\omega)$ be two extended Dirichlet series, both convergent somewhere. A type of automorphic pairs being given on G_A, let F be the function defined on B_A by the series (49) with the same coefficients as Z, and let F' be the function similarly derived from Z'. Then (F, F') can be extended to a B-cuspidal automorphic pair (Φ, Φ') of the given type if and only if, for all positive divisors $\mathfrak{n}$ disjoint from S, the Dirichlet series $Z_{\mathfrak{n}}$, $Z'_{\mathfrak{n}}$ (derived from Z, Z' by the Hecke operator $T_{\mathfrak{n}}$) can be continued as entire functions on Ω_S and satisfy the functional equation

$$G_\infty(\omega_\infty) Z_{\mathfrak{n}}(\omega)$$
$$= \omega_\infty(-1) e_\infty(\omega_\infty) \alpha(\mathfrak{f}_0 \mathfrak{n}) \omega(\mathfrak{a}) \varepsilon(\omega)^2 G_\infty(\alpha_\infty^{-1} \omega_\infty^{-1}) Z'_{\mathfrak{n}}(\omega^{-1}) \;,$$

the common value of both sides being bounded in each strip $A \leqq \sigma(\omega) \leqq B$ on each component of Ω_S.

The proof proceeds just as that of theorem 3 in Chapter VII, §32. Applying theorem 1 of Chapter IV, §18, we see that we have to verify (45) whenever $\mathfrak{f} = \operatorname{div} f$ is disjoint from S. As $k_A^1/k^\times$ is a compact group, this is the same as to verify (46) under the same assumption; as both sides are 0 when the conductor $\mathfrak{f}_0$ of ω does not divide $\mathfrak{f}$, we may assume that it does, and therefore that it is also disjoint from S. As all the calculations in Chapter VII, §30 are purely local ones at a finite place not occurring in $\mathfrak{A}$, they remain valid here, so that (19) and (20) of §30 are valid. We can now repeat the proof of lemma 5 in §31, merely replacing the conclusion of that lemma by our formula (46), §64, and modifying the assumption of the lemma accordingly. Thus all we have to do is to verify the analogue of (46) for $Z_{\mathfrak{n}}$, $Z'_{\mathfrak{n}}$, $\mathfrak{n}$ being any divisor disjoint from S, under the assumption $\mathfrak{f} = \mathfrak{f}_0$. This amounts to proving that the conclusion of theorem 6, concerning Z and Z', implies the validity of (46) whenever $\mathfrak{f} = \mathfrak{f}_0$.

Let again $\mathfrak{C}$ be a connected component of Ω_k with the conductor $\mathfrak{f}_0$, and choose any $\chi \in \mathfrak{C}$; then every quasicharacter in $\mathfrak{C}$ can be written as $\omega = \chi\omega_{s/n}$ with $s \in \mathbf{C}$. Note that, according to (42), $I(f, e, t, \omega)$ does not change if we replace ω by any other quasicharacter in the same connected component of Ω_k. We will now apply lemma 12 to the following functions. As function φ in that lemma, we take

$$\varphi(\nu) = I(f, e, tu_\nu, \chi)\chi(tu_\nu) \ .$$

For φ' in the lemma, we take the function similarly derived from the right-hand side of (46), i.e.:

$$\varphi'(\nu) = I'(f, e', at^{-1}u_\nu^{-1}, \chi^{-1})\chi(tu_\nu)M'_\infty(\mathfrak{p}_\infty)a_\infty(f_\infty)a(\mathfrak{f}_0) \ .$$

Then the integral $f(s)$ in the lemma is nothing else than $|t|^{-s/n}J(\chi\omega_{s/n})$, where $J(\omega)$ is again the integral (43) and can be evaluated by means of proposition 14 as in the proof of theorem 6. Quite similarly, the integral $f'(s)$ in the lemma can be expressed in terms of $J'(\chi^{-1}\omega_{-s/n})$ and can be evaluated in the same manner. With these notations, (46) and (47) are in substance nothing else than the equalities $\varphi = \varphi'$, $f = f'$, respectively. In the proof of theorem 6, we derived the functional equation for Z and Z' from (47), by combining it with the results of §58 (the "functional equation at infinity"). The same arguments, in reverse order, will show now that, if one assumes the functional equation for Z and Z', f and f' can be continued to one and the same entire function, so that our proof will be complete if we verify for f and f' the estimates required by lemma 13. Now it is obvious that $Z(\chi\omega_{s/n})$ is bounded for any fixed value of $\mathrm{Re}(s)$ for which it is absolutely convergent; combining this with the first part of lemma 13, we see that $f(\sigma + it)$ is $O(|t|^{-M})$ for all M, if σ is large enough; actually, the remarks following lemma 13 show that it is even $O(e^{-\delta|t|})$ for a suitable $\delta > 0$. The same argument, applied to Z' and Γ'_∞, shows the same for $f(\sigma' + it)$ for $-\sigma'$ large enough. Finally, the boundedness assumption for both sides of the functional equation, combined with the second part of lemma 13, implies that $f(s)$ is $O(e^{\gamma|t|})$ for some γ, with $t = \mathrm{Im}(s)$, uniformly within any given strip; then lemma 11, applied e.g. to $f(s)s^2$, gives the desired conclusion.

Corollary. *Assumptions being as in theorem 7, assume also that for every prime divisor $\mathfrak{p}$ not in S there is λ such that F and F' (or Z and Z') are eigenfunctions of the Hecke operator $T_{\mathfrak{p}}$ for the eigenvalues λ, $\lambda\alpha(\mathfrak{p})^{-1}$, respectively. Then theorem 7 remains valid if we restrict the condition in it to the case $\mathfrak{n} = 1$.*

This is obvious.

The above theorem and its corollary conclude our investigation. As at the end of Chapter VII, it is worth noting that the condition in the corollary is fulfilled if (and only if) the Dirichlet series Z, Z' are eulerian at all finite places $\mathfrak{p}$ not in S, with Euler factors of the form described in §§24-25 of Chapter VI. This is an immediate consequence of the results of those §§ and of the definition of Z and Z' in the present

Chapter. It is usually in this form that the condition in the corollary is applied.

CHAPTER XI

EXAMPLES

69. From the elementary point of view which we have adopted in these lectures, the only way of showing that our theory is not empty is to give examples. As appears from Chapters I and IX, the classical theory of modular functions already gives some; in fact, the modular forms (of arbitrary degree) belonging to the Hecke groups $\Gamma_o(N)$, as defined in §5, are in substance identical with those automorphic forms of our theory which belong to the field $k = \mathbb{Q}$ and to any discrete type on $G_\infty = GL(2, \mathbb{R})$. More generally, holomorphic automorphic forms for the Hilbert modular group over a totally real field k give such examples where the type, at each infinite place, is a discrete one. Principal types at real places were first considered by Maass in the case $k = \mathbb{Q}$; his work also showed the possibility of dealing with other than totally real fields by similar methods. In all those cases, the starting point is the automorphic form.

In this Chapter, we wish to discuss two kinds of examples, of special arithmetical significance, where the starting point is the Dirichlet series.

70. One class of examples consists of the zeta-functions of elliptic curves over the given ground-field k; they are defined as follows. Let E be an elliptic curve defined over k, i.e. a curve of genus 1 with at least one rational point; we exclude the trivial case where k is a function-field (over a finite field of constants k_o) and E is a "constant" curve, i.e. isomorphic over k to an elliptic curve defined over k_o.

Let v be a finite place of k; write $a \longrightarrow \bar{a}$ for the canonical isomorphism of the ring r_v onto the finite field $\mathfrak{p}_v = r_v/\pi_v r_v$ (the "residue field" at v), i.e. for the "reduction modulo $\mathfrak{q}_v$". Among the curves isomorphic to E over k_v, defined by an equation

$$Y^2 + \lambda XY + \mu Y = X^3 + \alpha X^2 + \beta X + \gamma$$

with coefficients in r_v, call E_v one whose discriminant has the lowest possible order. As Néron has shown, this is essentially unique, and the same equation "reduced modulo $\mathfrak{q}_v$" (i.e. with λ, etc., replaced by $\bar{\lambda}$, etc.) defines an irreducible curve $\bar{E}_v$ over $\mathfrak{p}_v$.

The place v is called a place "of good reduction" for E if $\bar{E}_v$ is an elliptic curve. Then the numerator of its zeta-function is of the form $1 - c_v T + q_v T^2$, where we put, as usual, $q_v = \mathrm{Card}(\mathfrak{p}_v) = |\mathfrak{q}_v|^{-1}$; c_v is an integer, and $1 - c_v + q_v$ is the number of rational points over $\mathfrak{p}_v$ on $\bar{E}_v$, including the point at infinity. That being so, we attach to the place v the Euler factor of degree 2:

$$Z_v(s) = (1 - c_v |\mathfrak{q}_v|^{1+s} + |\mathfrak{q}_v|^{1+2s})^{-1} .$$

Otherwise we put

$$Z_v(s) = (1 - \delta_v |\mathfrak{q}_v|^{1+s})^{-1}$$

where δ_v is ± 1 if $\bar{E}_v$ has a node and 0 if it has a cusp; in the former case, it is 1 if the two tangents at the node are rational over $\mathfrak{p}_v$, and -1 otherwise. Then we put

$$Z_E(s) = \prod_v Z_v(s) ,$$

where the product is taken over all the finite places of k. Because of the "Riemann hypothesis" for elliptic curves, we have $|c_v| \leqq 2q_v^{1/2}$ for all v, so that $Z_E(s)$ is a Dirichlet series belonging to k, absolutely convergent for $\mathrm{Re}(s) > 1/2$; it is called <u>the zeta-function of</u> E. If k is a function-field, k defines an algebraic curve C over its field of constants k_o, and E may be regarded as the "generic fiber" of an elliptic surface defined over k_o; in that case, Z_E is closely related to the so-called zeta-function of this surface over k_o.

One also attaches to E a positive divisor $\mathfrak{a}_E$, its "conductor"; it has the following properties. It contains a place $\mathfrak{p}$ of k if and only if it is not a place of good reduction; if then $\bar{E}_v$ has a node, the exponent of $\mathfrak{p}$ in $\mathfrak{a}_E$ is 1; if $\bar{E}_v$ has a cusp and $\mathfrak{p}_v$ is not of characteristic 2 or 3, that exponent is 2; otherwise it is $\geqq 2$, and its precise value can be defined either geometrically or by ramification properties which we will not discuss here. There is always a finite algebraic extension k' of k such that, over k', only nodes can occur.

On the basis of various examples, it has been conjectured that, for a suitable $\varepsilon = \pm 1$, the pair $Z = Z_E$, $Z' = \varepsilon Z_E$ satisfies all the conditions of our theorem 3 if k is a function-field and of our theorem 7 if k is a number-field, the type being defined by $\alpha = 1$, $\mathfrak{a} = \mathfrak{a}_E$, and by $\mathcal{H}_\infty$ on G_∞ in the latter case. This has now been proved by Deligne in the function-field case. Over number-fields, it has not yet been possible to deal with any curve except the following:

(a) Curves with complex multiplication, whose zeta-function has been determined by Deuring. If E is such a curve, and the complex multiplications are defined over k, Z_E is a product of two Hecke L-functions over k; if not, they define a quadratic extension k' of k, and Z_E is a Hecke L-function over k'.

(b) Various curves, related to certain arithmetic groups, for which the zeta-function has been determined, in some cases by Eichler and later in much more general cases by Shimura. We will only quote here one typical example (the first one to be treated by Eichler). If we write H for the Poincaré half-plane $\mathrm{Im}(\tau) > 0$, the quotient $H/\Gamma_o(11)$, compactified in the usual manner, defines a Riemann surface of genus 1; it has models, defined over $\mathbb{Q}$; one such model, given by Fricke, is

$$Y^2 = 1 - 20X + 56X^2 - 44X^3 \quad ;$$

as Tate has observed, this is isogenous to $Y^2 - Y = X^3 - X^2$. The conductor is 11; the zeta-function is the Mellin transform of the (unique) cusp-form of degree -2 belonging to $\Gamma_o(11)$.

It seems possible that this last example is typical, in the following sense. Assume the truth of the conjecture stated above, for some elliptic curve E over $k = \mathbb{Q}$; let Φ be the automorphic form of type $(1, \mathfrak{n}_E, \mathcal{H}_R)$ corresponding to the Dirichlet series Z_E; if we put $N = \mathfrak{n}_E$, this is essentially the same (according to our Chapter I) as a cusp-form $f(\tau)$ of degree -2 belonging to the group $\Gamma_o(N)$, i.e. as a differential $f(\tau)d\tau$ of the first kind belonging to the Riemann surface $H/\Gamma_o(N)$. It happens in some cases, and it might be true in general, that $f(\tau)d\tau$ has only two linearly independent periods over $\mathbf{Z}$, so that it defines an elliptic curve E' over $\mathbb{C}$. Under such circumstances, as Shimura has observed, E' has a model defined over $\mathbb{Q}$. Then the question arises: is this model isogenous to the original curve E?

Similar questions arise, in vaguer form, for curves over algebraic number-fields $k \neq \mathbb{Q}$, if one still assumes the truth of the same conjecture. As we have seen in Chapter X, §57, one can associate with any admissible function on G_∞ a harmonic differential form of degree r (the number of infinite places of k), and another one of degree $n = [k : \mathbb{Q}]$, on the Riemannian symmetric space defined by G_∞. With any automorphic form of type $(1, \mathfrak{n}_E, \mathcal{H}_\infty)$ on G_A, one can thus associate one or (in general) more than one such differential form, invariant under certain arithmetic groups. We are now led to ask: if this is so, is there a relation between the periods of those forms and the periods of the elliptic curve E and its conjugates over $\mathbb{Q}$?

71. The examples discussed above under (a) may also be

regarded as belonging to another class, the "Artin-Hecke L-functions". We will now describe briefly how these are defined.

To every A-field k, one attaches canonically a group $\mathfrak{G}_k$ with the following properties: (a) it is "quasicompact"; more precisely, it is the direct product of a group isomorphic to $\mathbf{R}$ and a compact group, or the semidirect product of a group isomorphic to $\mathbf{Z}$ and a compact group, according as the characteristic of k is 0 or not; (b) if $\mathfrak{G}_k^{(c)}$ is the closure of the commutator subgroup of $\mathfrak{G}_k$, $\mathfrak{G}_k/\mathfrak{G}_k^{(c)}$ is canonically isomorphic to the idele-class group $k_A^\times/k^\times$ with which we will frequently identify it; (c) there is a one-to-one correspondence "of Galois type" between the separable extensions k' of k of finite degree d (within a fixed separable algebraic closure k_{sep} of k) and the subgroups of $\mathfrak{G}_k$ of index d; moreover, the subgroup corresponding to k' is canonically isomorphic to $\mathfrak{G}_{k'}$ with which it will be identified; (d) if k' is a finite Galois extension of k, $\mathfrak{G}_k/\mathfrak{G}_{k'}$ is canonically isomorphic to the Galois group of k' over k. If k is a function-field, $k_o = \mathbf{F}_q$ its field of constants, and $\bar{k}_o$ the algebraic closure of k_o, $\mathfrak{G}_k$ is nothing else than the group of those automorphisms of k_{sep} over k which induce on $\bar{k}_o$ some power of $x \longrightarrow x^q$; it is topologized so that its maximal compact subgroup is the Galois group of k_{sep} over $\bar{k}_o k$ with its usual topology; thus, in this case, the description of $\mathfrak{G}_k$ requires nothing more than Galois theory combined with the classfield theory for k. If k is a number field, the quotient group of $\mathfrak{G}_k$ by the connected component of the neutral element ε is the Galois group of $\bar{k} = k_{sep}$ over k; for the details of the construction of $\mathfrak{G}_k$ in that case, cf. A. Weil, J. Math. Soc. Japan 3 (1951), pp. 1-35.

72. As may easily be seen, every irreducible representation M of $\mathfrak{G}_k$ is of finite degree (but not necessarily equivalent to a unitary representation); it is uniquely characterized by its trace χ, also known as its "character"; its degree is $n = \chi(\varepsilon)$. When $n = 1$, the

kernel of M contains $\mathfrak{G}_k^{(c)}$, so that both M and χ may be identified with a quasicharacter of the group $\mathfrak{G}_k/\mathfrak{G}_k^{(c)}$, i.e. of $k_A^\times/k^\times$; in particular, any quasicharacter of the latter group of the form ω_s, with $s \in \mathbf{C}$, may be so regarded. For any n, the kernel of M is contained in $\mathfrak{G}_{k'}^{(c)}$ for some finite Galois extension k' of k; thus, when only finitely many irreducible representations are to be considered, one may, instead of $\mathfrak{G}_k$, consider a suitable quotient $\mathfrak{G}_k/\mathfrak{G}_{k'}^{(c)}$; the latter group is sometimes denoted by $\mathfrak{G}_{k',k}$. The trace χ of any representation M of $\mathfrak{G}_k$ of finite degree may be written as $\Sigma a_i \chi_i$, where the χ_i are traces of irreducible representations and the a_i are integers > 0; then the degree of M is again $\chi(\varepsilon)$. In particular, if χ, χ' are the traces of two such representations, the trace $\chi\chi'$ of $M \otimes M'$ can be written in that form. Thus the functions on $\mathfrak{G}_k$ of the form $\Sigma a_i \chi_i$, where the χ_i are the traces of non-equivalent irreducible representations and the a_i are now any integers, make up a ring X_k, whose elements will be called <u>the characters</u> of $\mathfrak{G}_k$; sometimes they are called "virtual characters", to emphasize the fact that they are not necessarily traces of representations (they are so only if all a_i are $\geqq 0$); we will rather say that a character is <u>positive</u> if it is the trace of a representation, i.e. if all a_i are $\geqq 0$. By a <u>prime character</u>, we will understand the trace of an irreducible representation. If χ is any character, the integer $\chi(\varepsilon)$ will be called its <u>degree</u>.

For any representation M of $\mathfrak{G}_k$, the determinant det M is a representation of $\mathfrak{G}_k$ of degree 1, which clearly depends only upon the character χ of M, and for which we write δ_χ; one sees at once that $\chi \longrightarrow \delta_\chi$ can be extended to a morphism of the additive group of X_k into Ω_k.

Let α be any quasicharacter of $k_A^\times/k^\times$, or, what amounts to the same, any prime character of $\mathfrak{G}_k$ of degree 1. Then, if χ is a character of degree n, so is $\chi\alpha$; if χ is prime (resp. positive), so is $\chi\alpha$. On X_k, we will introduce a topology and a complex structure, similar to those introduced on Ω_k in §9, by agreeing that the connected component of any character χ consists of all the characters of the form $\chi\omega_s$, and that s defines the complex structure on that component.

Let k' be any separable extension of k of finite degree d; then $\mathfrak{G}_{k'}$ is of index d in $\mathfrak{G}_k$. Let M' be any representation of $\mathfrak{G}_{k'}$ of finite degree n; from this, one derives in a well-known manner a representation M of $\mathfrak{G}_k$ of degree nd, traditionally known as the one induced by M'; the trace χ of M depends only upon k, k' and the trace χ' of M'; it is said to be induced by χ' and will be denoted by $[\chi'; k'/k]$. The mapping $\chi' \longrightarrow [\chi'; k'/k]$ can then be extended to a morphism of the additive group of $X_{k'}$ into that of X_k. We will say that $[\alpha'; k'/k]$ is an elementary character of $\mathfrak{G}_k$ if α' is a prime character of degree 1 of $\mathfrak{G}_{k'}$, i.e. a quasicharacter of $k'^{\times}_A/k'^{\times}$; here one could, without affecting the validity of what follows, modify this definition by imposing further restrictions on k', but this is not needed for our purposes. By using Brauer's theorem, one proves (cf. loc. cit.) that the elementary characters generate the additive group of X_k.

73. In §6, we gave the definition of a Dirichlet series belonging to k; here we will consider only those whose initial coefficient c(1) is not 0; this will be tacitly assumed from now on. If k' is any finite extension of k, let

$$L' = \sum c'(\mathfrak{m}')\,|\mathfrak{m}'|^{s}$$

be a Dirichlet series belonging to k'; from it, we derive a Dirichlet series L belonging to k, viz., the one with the coefficients

$$c(\mathfrak{m}) = \sum_{N\mathfrak{m}'=\mathfrak{m}} c'(\mathfrak{m}') \quad ;$$

here the summation is taken over all positive divisors $\mathfrak{m}'$ of k' whose relative norm $N_{k'/k}(\mathfrak{m}')$ is equal to $\mathfrak{m}$. Then the function L(s) defined in the s-plane by the latter series coincides with the function L'(s) defined by the former one. More generally, write $L'(\omega')$, $L(\omega)$ for the "extended Dirichlet series" defined respectively on $\Omega_{k'}$ by L' and on Ω_k by L, in the manner explained in §9. The norm mapping $N_{k'/k}$ defines a morphism of $k'^{\times}_A/k'^{\times}$ into $k^{\times}_A/k^{\times}$, and therefore a morphism $\omega \longrightarrow \omega \circ N_{k'/k}$ of Ω_k into $\Omega_{k'}$; it is then easily seen that $L(\omega)$ is nothing else than $L'(\omega \circ N_{k'/k})$. We will write $[L'; k'/k]$ for the series L.

To each k' and to every character $\chi' \in X_{k'}$, we attach now a divisor $\mathfrak{f}_{k'}(\chi')$ of k' and a Dirichlet series $L_{k'}(\chi')$ belonging to k', so as to satisfy the following conditions:

(I) If α' is a prime character of degree 1 of $\mathfrak{G}_{k'}$, i.e. a quasicharacter of $k'^{\times}_A/k'^{\times}$, then $\mathfrak{f}_{k'}(\alpha')$ is its conductor, and $L_{k'}(\alpha')$ is the L-series for α' (as defined in §8).

(II) $\chi' \longrightarrow \mathfrak{f}_{k'}(\chi')$ and $\chi' \longrightarrow L_{k'}(\chi')$ are morphisms of the additive group of $X_{k'}$ into the multiplicative groups of the divisors of k', and of the Dirichlet series belonging to k' with the initial coefficient 1, respectively.

(III) Let k', k'' be any two finite separable extensions of k, such that $k \subset k' \subset k''$; let $D(k''/k')$ be the discriminant of k'' over k'; let χ'' be any character of $\mathfrak{G}_{k''}$, and let $n = \chi''(\varepsilon)$ be its degree. Then:

$$\mathfrak{f}_{k'}([\chi''; k''/k']) = N_{k''/k'}[\mathfrak{f}_{k''}(\chi'')]D(k''/k')^n$$

$$L_{k'}([\chi''; k''/k']) = [L_{k''}(\chi''); k''/k'] .$$

For brevity's sake, we will usually write $\mathfrak{f}(\chi')$, $L(\chi')$ instead of $\mathfrak{f}_{k'}(\chi')$, $L_{k'}(\chi')$; and we will write, whenever convenient, $\mathfrak{f}_{\chi'}$ instead of $\mathfrak{f}(\chi')$. Clearly (I) and (III) together determine $\mathfrak{f}(\chi')$, $L(\chi')$ for all elementary characters of $\mathfrak{G}_{k'}$; as the latter generate $X_{k'}$, our conditions determine these uniquely for all characters, if at all, i.e. if they can be fulfilled. That this can be done (at least for those characters which can be defined by Galois theory, i.e. those belonging to representations which are trivial on the connected components of the groups $\mathfrak{G}_{k'}$) was proved by Artin in some classical papers and remains one of his major achievements. For the extension to the general case, which offers no serious difficulty once the groups $\mathfrak{G}_k$ have been constructed, cf. A. Weil, loc. cit. One may remark that Brauer's theorem was originally not available to Artin, who had to replace it by a somewhat weaker substitute, with correspondingly weaker results.

Since obviously all that can be said about $\mathfrak{G}_k$ remains valid for $\mathfrak{G}_{k'}$, we will mostly confine our statements to $\mathfrak{G}_k$ and its characters. The case of a positive character χ deserves special consideration; in that case, as Artin showed, $\mathfrak{f}_\chi$ is a positive divisor, known as the conductor of χ, and $L(\chi)$ is eulerian of degree $n = \chi(\varepsilon)$ at all finite places not in $\mathfrak{f}_\chi$, and of degree $< n$ at the places occurring in $\mathfrak{f}_\chi$. More precisely, one defines for each place v of k, finite or not, a class of subgroups $\mathfrak{G}_{k,v}$ of $\mathfrak{G}_k$ (the "decomposition groups" for v), conjugate to each other under the inner automorphisms of $\mathfrak{G}_k$. Any one of these groups is the analogue for k_v of the group $\mathfrak{G}_k$ for k and is the semidirect product of its maximal compact subgroup $\mathfrak{G}^1_{k,v}$ and of a group isomorphic to $\mathbf{R}$ if v is infinite, to $\mathbf{Z}$ otherwise. Then, for the character χ of any representation M of $\mathfrak{G}_{k'}$, the order of $\mathfrak{f}_\chi$ and the Euler factor of $L(\chi)$ at a finite place v depend only upon the restriction of M to $\mathfrak{G}_{k,v}$; in particular: (a) if M is fully reducible, it is unramified at $\mathfrak{p}_v$ (i.e., $\mathfrak{p}_v$ does not occur in $\mathfrak{f}_\chi$) if and only if it is trivial on the "inertia group" $\mathfrak{G}^1_{k,v}$; (b) if it is so, and if F_v is a representative in $\mathfrak{G}_{k,v}$ of the generator of $\mathfrak{G}_{k,v}/\mathfrak{G}^1_{k,v}$ (a "Frobenius element") for which $\omega_1(F_v) < 1$, then the

Euler factor for $L(\chi)$ at $\mathfrak{p}_v$ is

(50) $$\det(1_n - M(F_v)|\mathfrak{p}_v|^s)^{-1}$$

with $n = \chi(\varepsilon)$. In particular, this must be so for $n = 1$, so that we have $\alpha(F_v) = \alpha(\mathfrak{p}_v)$ for every quasicharacter α of $k_A^\times/k^\times$, unramified at $\mathfrak{p}_v$; in other words, the image of F_v in $\mathfrak{G}_k/\mathfrak{G}_k^{(c)}$ (when the latter group is identified with $k_A^\times/k^\times$) is a prime element π_v of k_v (and the images of $\mathfrak{G}_{k,v}$, $\mathfrak{G}_{k,v}^1$ are $k_v^\times$, $r_v^\times$, respectively). For instance, $\omega_1(F_v) = |\mathfrak{p}_v|$. It also follows from (b) that $\delta_\chi = \det M$ is unramified at $\mathfrak{p}_v$ if M is so, and that we have then $\det M(F_v) = \delta_\chi(\mathfrak{p}_v)$.

74. So far, we have considered the series $L(\chi)$ as formal Dirichlet series; if now we consider the functions $L(\chi, s)$ which they define in the s-plane, it is clear, from (I), (II), (III) and Brauer's theorem that they are meromorphic there, and that each of them is holomorphic in some half-plane $Re(s) > \sigma$. Artin conjectured that, for every positive χ except $\chi = 1$, $L(\chi)$ is an entire function. For number-fields, this is still an open question, and presumably a formidable one. For function-fields, it was proved along with the Riemann hypothesis.

More generally, we can consider the "extended Dirichlet series" $[L(\chi)](\omega) = L(\chi, \omega)$ as functions on Ω_k, according to the definitions in §9; as above, we see that $\omega \longrightarrow L(\chi, \omega)$ is meromorphic on Ω_k for every χ. We will now show that we have, for all $\chi \in X_k$, all $\omega \in \Omega_k$, and all $s \in \mathbb{C}$:

(51) $$L(\chi, \omega\omega_s) = L(\chi\omega, s) .$$

This could be deduced from the Euler product for $L(\chi)$, as given by Artin; but it is simpler to observe that $\chi \longrightarrow L(\chi, \omega\omega_s)$ and $\chi \longrightarrow L(\chi\omega, s)$, for any fixed ω, are morphisms of the additive group of X_k into the multiplicative group of meromorphic functions in the s-plane. Therefore, by Brauer's theorem, it is enough to prove (51) for an elementary character $\chi = [\alpha'; k'/k]$, in which case, by (III), $L(\chi)$ can be written as $[L(\alpha'); k'/k]$, where $L(\alpha')$ is an "ordinary" L-series. Then the definition of an induced representation, combined with the basic facts of classfield theory, shows that $\chi\omega$ is also an elementary character, viz., the one induced by $\alpha'.(\omega \circ N_{k'/k})$, and (51) can be verified at once.

In substance, this shows that, for any fixed s_o, $\chi \longrightarrow L(\chi, s_o)$ is a meromorphic function on X_k, for the complex structure defined in §72, and that its knowledge implies that of all the formal Dirichlet series $L(\chi)$. For reasons arising from the special role of $s = 1/2$ in the functional equation of the ordinary L-series, we choose $s_o = 1/2$; consequently, from now on, we will write $L_o(\chi) = L(\chi, 1/2)$; this may take the values 0 and ∞. Then $L(\chi, s)$ is the same as $L_o(\chi\omega_{s-(1/2)})$, and the "extended Dirichlet series" $L(\chi, \omega)$ the same as $L_o(\chi\omega_{-1/2}\omega)$. For a positive χ of degree n, (50) shows that $L_o(\chi\omega_s)$ has at every place $\mathfrak{y}_v$ not in $\mathfrak{f}(\chi)$ the Euler factor

$$\det(1_n - M(F_v)|\mathfrak{y}_v|^{s+(1/2)}) .$$

As noted above, we have $\det M(F_v) = \delta_\chi(\mathfrak{y}_v)$. In particular, for $n = 2$, this shows that the Euler factor has the form indicated in Chapter VI, §24, with $\alpha = \delta_\chi$; thus, in that case, it seems worthwhile to investigate the Dirichlet series $Z(\omega) = L_o(\chi\omega)$ from the point of view of our Chapters VII resp. X, and try to apply to it our theorem 3 resp. 7. Clearly this must depend in the first place upon the setting up of a functional equation for such series, as will be done now for an arbitrary d.

75. If M is a representation of $\mathfrak{G}_k$ with the trace χ, we associate with it the contragredient representation ${}^tM^{-1}$, whose trace we will denote by $\hat{\chi}$; as a function on $\mathfrak{G}_k$, this is given by $\hat{\chi}(\gamma) = \chi(\gamma^{-1})$ for all $\gamma \in \mathfrak{G}_k$; for a prime χ of degree 1, we have $\hat{\chi} = \chi^{-1}$. For an induced character $\chi = [\chi'; k'/k]$, we have $\hat{\chi} = [\hat{\chi}'; k'/k]$.

For a function-field k, we will now prove the functional equation in the following form:

Lemma A. Let k be a function-field, and d_k a differental idele of k. There is a morphism η of the additive group X_k into $\mathbf{C}^\times$ such that, for all χ and all $s \in \mathbf{C}$:

$$L_o(\chi) = \eta(\chi)L_o(\hat{\chi}), \quad \eta(\chi\omega_s) = \eta(\chi)|\mathfrak{f}_\chi|^s \cdot |d_k|^{s\chi(\varepsilon)} .$$

Consider first an elementary character $\chi = [\alpha'; k'/k]$, where α' is a quasicharacter of $k'^\times_A/k'^\times$; then $L(\chi)$ is the same as the "ordinary" L-function $L_{k'}(\alpha')$, whose well-known functional equation can now be re-written in the form

$$(52) \qquad L_o(\chi\omega_s) = \varepsilon(\alpha')|\mathfrak{f}_\chi|^s \cdot |d_k|^{s\chi(\varepsilon)} L_o(\hat{\chi}\omega_{-s}) ,$$

where $\varepsilon(\alpha')$ is as before (cf. §10 and §27) and independent of s. Take then $\eta(\chi) = \varepsilon(\alpha')$; this defines a mapping η of the set of all elementary characters into $\mathbf{C}^\times$. Observe that in (52) all terms except $\varepsilon(\alpha')$, if re-garded as functions of χ, define morphisms of X_k into the multiplicative group of the meromorphic functions of s in the s-plane. This implies that can be uniquely extended to a morphism of X_k into $\mathbf{C}^\times$. Replacing χ by $\chi\omega_t$ in (52), with any $t \in \mathbf{C}$, we get the second formula in our lemma.

Lemma B (Langlands). Let η be as in lemma A; then, for any two positive characters χ, χ' of respective degrees n, n' with disjoint conductors $\mathfrak{f}_\chi$, $\mathfrak{f}_{\chi'}$, we have:

$$\eta(\chi\chi') = \delta_\chi(\mathfrak{f}_{\chi'})\delta_{\chi'}(\mathfrak{f}_\chi)\eta(\chi)^{n'}\eta(\chi')^n .$$

Calling $\bar{\eta}(\chi, \chi')$ the right-hand side for any χ, χ' and any morphism η of X_k into $\mathbf{C}^\times$, we will write $H(\chi, \chi') = \eta(\chi\chi')^{-1}\bar{\eta}(\chi, \chi')$; this is symmetric in χ, χ', and "bilinear", in the sense that $\chi' \longrightarrow H(\chi, \chi')$, for a fixed χ, is a morphism of X_k into $\mathbf{C}^\times$. As we have seen above, if S is any set of places of k, the condition for a fully reducible representation M of $\mathfrak{G}_k$ to be unramified at all places of S is that it should be trivial on certain subgroups of $\mathfrak{G}_k$, and therefore on the smallest closed normal subgroup G_S containing them all. Applying Brauer's theorem to the representations of $\mathfrak{G}_k/G_S$ when S is the set of all places not in the conductor $\mathfrak{f}_\chi$ of a given positive character χ, we see that χ can be expressed in terms of elementary characters which are unramified outside $\mathfrak{f}_\chi$. Thus, in order to prove the lemma, it will be enough to show that $H(\chi, \chi') = 1$ for η as in lemma A and for any two elementary characters χ, χ' whose conductors are disjoint.

As we have seen in §10, if α is a prime character of degree 1, we have $\eta(\alpha) = \kappa(\alpha)\alpha(d_k f_\alpha)$, where $\kappa(\alpha)$ is a product of local factors $\kappa_v(\alpha)$ and f_α is an idele such that $\mathrm{div}(f_\alpha) = \mathfrak{f}_\alpha$; the idele f_α enters into the definition of $\kappa(\alpha)$. In particular, we have $\eta(1) = 1$ and more generally $\eta(\omega_s) = |d_k|^s$, this being in substance the functional equation for the zeta-function of k. If α, β are two prime characters of degree 1, with disjoint conductors $\mathfrak{f}_\alpha, \mathfrak{f}_\beta$, the conductor of $\alpha\beta$ is $\mathfrak{f}_\alpha\mathfrak{f}_\beta$; we may take $f_{\alpha\beta} = f_\alpha f_\beta$, with e.g. $(f_\alpha)_v = 1$ at all places not in $\mathfrak{f}_\alpha$, and similarly for f_β. Then one sees at once that $\kappa_v(\alpha\beta)$ is the same as $\kappa_v(\alpha)$ at all places in $\mathfrak{f}_\alpha$, the same as $\kappa_v(\beta)$ at all places in $\mathfrak{f}_\beta$, and 1 everywhere else. This proves $H(\alpha, \beta) = 1$.

In what follows, whenever we write a symbol $H(\chi, \chi')$ with positive characters χ, χ', it will be understood that χ, χ' have disjoint conductors. We proceed step by step, using the fact that, if $\chi_1 \in X_k$, $\chi' \in X_{k'}$ and χ'_1 is the restriction of χ_1 to $\mathfrak{G}_{k'}$, then $\chi_1 \cdot [\chi'; k'/k]$ is the same as $[\chi'_1\chi'; k'/k]$.

a) We first prove $H(\alpha, [1; k'/k]) = 1$, for a prime α of degree 1; call α' the restriction of α to $\mathfrak{G}_{k'}$, which, regarded as a quasi-character of $k'^\times_A/k'^\times$, is the same as $\alpha \circ N_{k'/k}$. Then $\eta(\alpha) = \kappa(\alpha)\alpha(d_k f_\alpha)$. As k' is unramified over k at all places in $\mathfrak{f}_\alpha$, we may take $f_{\alpha'} = f_\alpha$, so that $\eta(\mathfrak{a}') = \kappa(\mathfrak{a}')\mathfrak{a}'(d_{k'}f_{\mathfrak{a}})$. Using the fact that $D_{k'/k} = \mathrm{div}[N_{k'/k}(d_k^{-1}d_{k'})]$, one sees that the result to be proved can be localized as follows. Let w be a place of k' above a place v in $\mathfrak{f}_{\mathfrak{a}}$; let ν be the degree of k'_w over k_v, m the order of $f_{\mathfrak{a}}$ at v; then $\kappa_w(\mathfrak{a}'_w) = \kappa_v(\mathfrak{a}_v)^{\nu}(-1)^{m(\nu-1)}$. For $m = 1$ (in which case it is essentially the same as the Hasse-Davenport theorem), a simple proof will be found in A. Weil, Bull. Am. Math. Soc. 55 (1949), on pp. 503-505. A similar proof can be given for any odd $m = 2\mu + 1$ with $\mu > 0$, by observing that, for a suitable choice of $a \in r_v^\times$, the function $u \longrightarrow \mathfrak{a}_v(1 + \pi_v^{\mu}u)\psi_v(d_v^{-1}f_v^{-1}\pi_v^{\mu}au)$ on r_v depends only upon the value of u modulo $\mathfrak{p}_v$ and may therefore be regarded as a mapping f of the residue field $\mathfrak{k}_v = r_v/\mathfrak{p}_v$ into $\mathbf{C}$; so is $u \longrightarrow \psi_v(-d_v^{-1}f_v^{-1}\pi_v^{2\mu}au)$, which is a non-trivial additive character of $\mathfrak{k}_v$ for which we write g; on page 503 of the proof just quoted, change the choice of $\lambda(F)$ to $\lambda(F) = f(c_1)g(c_2)$; then the proof proceeds just as there. Finally, for an even $m = 2\mu$, the "Gaussian sum" $\kappa_v(\mathfrak{a}_v)$ turns out (by an elementary calculation) to be $\mathfrak{a}_v(a)\psi_v(d_v^{-1}f_v^{-1}a)$, with a so chosen that the function $f(u)$ defined as above is 1 for all u. Then one can choose the same a for $\mathfrak{a}_v$ in k_v and for $\mathfrak{a}'_w$ in k_w, and the result follows trivially, provided we have chosen $\psi' = \psi \circ \mathrm{Tr}_{k'/k}$, as has also been tacitly assumed above.

b) From here on, the proof proceeds on purely formal lines. We will write $\Delta_{k'}$ for the character, equal to ± 1, given by $\delta_{[1;k'/k]}$. As a consequence of the "transfer theorem" of classfield theory, we have, for any quasicharacter β' of $k'^\times_A/k'^\times$, $\delta_{[\beta';k'/k]} = \Delta_{k'}\cdot(\beta' \circ \mathrm{inj})$, where inj is the natural injection of $k_A^\times/k^\times$ into $k'^\times_A/k'^\times$. Now, for any such β', let χ be a positive character of $\mathfrak{J}_k$ with a conductor disjoint from that of $[\beta'; k'/k]$ (as given by (III) of §73); let χ' be the restriction of χ to $\mathfrak{J}_{k'}$. A formal calculation gives the formula

$$H(\chi, [\beta'; k'/k]) = H_{k'}(\chi', \beta').H(\chi, [1, k'/k]) \ , \qquad (53)$$

where $H_{k'}$ is the symbol, similar to H but taken over k'.

c) In (53), take $\chi = \alpha$, with α prime of degree 1. Then we already know that the two factors in the right-hand side are 1, so that also the left-hand side is 1. In view of Brauer's theorem, this proves $H(\alpha, \chi) = 1$ for all χ (under the same assumptions as before). This case would actually suffice for the application to the characters of degree 2 which was our main motivation for taking up these questions.

d) In (53), we know now that the first factor in the right-hand side is always 1. By Brauer's theorem, our proof will be complete if we prove $H(\chi, [1; k'/k]) = 1$; applying the same argument to χ in the latter symbol, and making again use of (53) (where we now substitute $[1; k', k]$ for χ, and, say, β'' and k'' for β' and k'), we see that it only remains for us to treat $H([1, k'/k], [1, k''/k])$, for which we will write $H(k', k'')$; here the discriminants $D' = D(k'/k)$, $D'' = D(k''/k)$ are assumed to be disjoint. We have now

$$H(k', k'') = \Delta_{k'}(D'')\Delta_{k''}(D') \quad ,$$

and apply to this an argument communicated by J.-P. Serre. Call $\mathfrak{f}', \mathfrak{f}''$ the conductors of $\Delta_{k'}, \Delta_{k''}$. If L is any quadratic extension of k, and α the corresponding character (of order 2), we have $\alpha = [1; L/k] - 1$, hence $\eta(\alpha) = 1$. As we know that $H(\Delta_{k'}, \Delta_{k''}) = 1$, we conclude that $\Delta_{k'}(\mathfrak{f}'')\Delta_{k''}(\mathfrak{f}') = 1$; therefore it will be enough to show, e.g. for k', that D' and $\mathfrak{f}'$ differ only by a square. Let $(\xi_1, \ldots, \xi_n)$ be any basis of k' over k; put $\delta = \det \mathrm{Tr}(\xi_i \xi_j)$; it is well-known that D' differs by a square from $\mathrm{div}(\delta)$. In particular, for $k' = k(\xi)$, apply this to the basis $(1, \xi, \ldots, \xi^{n-1})$; then, if $\sigma_1, \ldots, \sigma_n$ are the distinct isomorphisms of k' into an algebraic closure $\bar{k}$ of k, we have $\delta = \delta_o^2$, with

$$\delta_o = \prod_{i<j} [\sigma_i(\xi) - \sigma_j(\xi)] \quad .$$

If the characteristic of k is 2, δ_o, being separable over k and invariant under all automorphisms of $\bar{k}$ over k, is in k; this shows that δ, hence also D', are squares; applied to a quadratic extension of k, this shows that the conductor $\mathfrak{f}_a$ of any character of order 2 is a square; so $\mathfrak{f}'$ is also a square. If the characteristic is not 2, $k(\delta_o)$ is the extension of k (of degree 1 or 2) corresponding to the character $\Delta_{k'}$; then $\mathfrak{f}'$, which is the discriminant of this extension, differs from $\mathrm{div}((2\delta_o)^2)$, hence from $\mathrm{div}(\delta)$, by a square. This completes the proof of lemma B.

76. Let now χ be any positive character of degree 2 of $\mathfrak{G}_k$. Define two Dirichlet series Z, Z' by

$$Z(s) = L_o(\chi\omega_s)\ , \qquad Z'(s) = \eta(\chi)L_o(\hat{\chi}\omega_s)\ ; \tag{54}$$

the corresponding extended Dirichlet series are

$$Z(\omega) = L_o(\chi\omega)\ , \qquad Z'(\omega) = \eta(\chi)L_o(\hat{\chi}\omega)\ .$$

Using (50), one verifies at once that Z, Z' are eulerian at all places of k outside $\mathfrak{f}_\chi$; their Euler factors have the form prescribed by §§24-25, with $a = \delta_\chi$ and with $\lambda = |\mathfrak{p}_v|^{-1/2}\mathrm{tr}\, M(F_v)$ at the place $\mathfrak{p}_v$. Using lemmas A and B, one verifies at once that they satisfy the functional equation of theorems 2 and 3, Chapter VII, with a as we have just said and $\mathfrak{A} = \mathfrak{f}_\chi$, for all the quasicharacters ω whose conductor is disjoint from $\mathfrak{A}$.

We can now apply theorem 3 of Chapter VII, or rather its corollary, to Z and Z', provided $Z(\omega)$ and $Z'(\omega)$ are holomorphic as required by that theorem. As to that, we have already mentioned that all Artin L-functions are known to be everywhere holomorphic, except for the zeta-function of k which has the poles ω_o, ω_1. Therefore, if χ is prime, i.e. if it is the character of an irreducible representation of degree 2, this condition is satisfied. Otherwise, we have $\chi = \alpha_1 + \alpha_2$, where α_1, α_2 are quasicharacters of $k_A^\times/k^\times$; call $\mathfrak{f}_1, \mathfrak{f}_2$ their conductors, so that $\mathfrak{f}_\chi = \mathfrak{f}_1\mathfrak{f}_2$. The application of theorem 3 requires here that $L(\alpha_1\omega)$ and $L(\alpha_2\omega)$ should be holomorphic whenever the conductor of ω is disjoint from $\mathfrak{f}_1\mathfrak{f}_2$; this is clearly the case unless $\mathfrak{f}_1$ or $\mathfrak{f}_2$ is 1. Thus we have proved the following:

Theorem. *Let* k *be a function-field; let* χ *be a positive character of degree* 2 *of* $\mathfrak{G}_k$. *Then the Artin* L-*series* Z, Z' *given by* (54) *are those attached to a* B-*cuspidal pair of automorphic functions* Φ, Φ', *belonging to the quasicharacter* δ_χ *of* $k_A^\times/k^\times$ *and to the conductor* $\mathfrak{f}_\chi$, *unless* χ *splits into two characters of degree* 1, *and at least one of these is unramified.*

The latter case, which we have been compelled to leave out, could be treated by the same method, had we given in Chapter VII the converse of proposition 8; had we done that, it would appear that Z and Z' still belong to an automorphic pair Φ, Φ', but Φ or Φ' (or both) would fail to be B-cuspidal. It has been shown by Jacquet and Langlands that Φ and Φ' are both cuspidal (in the sense explained in §27) if and only if χ is prime, i.e. belongs to an irreducible representation, and that otherwise they are "Eisenstein series".

77. It is noteworthy that, whenever E is a "twisted constant curve" over the function-field k (i.e., one which is not constant but has a constant absolute invariant $j(E)$, so that it becomes constant over some finite extension of k), its zeta-function is an L-function $L(\chi)$ belonging to a positive character χ of degree 2 of $\mathfrak{G}_k$. Here χ splits into two prime characters of degree 1 if there is a cyclic extension of k over which E becomes constant; this is so e.g. if the characteristic p is not 2 and E is given by an equation $wY^2 = X^3 - aX - b$, with a, b in the field of constants k_o of k, w being in $k^\times$ but not in $k_o^\times(k^\times)^2$. Otherwise χ is prime; then $j(E)$

must be 0 or 12^3; if at the same time p is not 2 or 3, χ (if it is prime) must be of the form $[\alpha'; k'/k]$, where k' is the compositum of k with the field $\mathbf{F}_q$ with $q = p^2$ elements and is (under that same assumption) of degree 2 over k, and α' is a prime character of degree 1 for k'; in this case, $L(\chi)$ is still an "ordinary" L-function, but one for k'. The most interesting cases occur for $p = 3$ and $p = 2$; then it may happen that the smallest Galois extension of k over which E becomes constant is not abelian (Galois groups of order 12 resp. 24 can occur); when that is so, $L(\chi)$ is not an "ordinary" L-function.

The results of §§71-75, and more particularly lemma B, also play an essential part in Deligne's treatment of the zeta-function of an arbitrary elliptic curve over a function-field. Actually, this zeta-function may be regarded as a kind of Artin L-function, attached, not to a representation of $\mathfrak{G}_k$ into $GL(2, \mathbf{C})$ as before, but to one into $GL(2, \mathbf{Q}_\ell)$, where ℓ is any prime other than the characteristic of k. Some very general results of Grothendieck's imply that it is holomorphic and satisfies a functional equation of the desired form. Those results say nothing about the "constant factor" in this equation; they lend themselves, however, to "reduction modulo ℓ". After this reduction, one has to do with a representation of $\mathfrak{G}_k$ into $GL(2, \mathbf{F}_\ell)$, whose character, because of Brauer's theory of "modular" representations, can be lifted back to a character (not necessarily positive) of $\mathfrak{G}_k$ in the sense of §72. To this, one can apply lemma B. Thus one sees that the "constant factor" in Grothendieck's functional equation has all the required properties modulo ℓ; as this is true for almost all ℓ, the conclusion follows. For more information on this subject, the reader may consult J.-P. Serre, <u>Facteurs locaux</u>, etc., Séminaire Delange-Pisot-Poitou, 11e année (1969/70), n° 19 (where he will find an extensive bibliography), and P. Deligne, <u>Les constantes</u>, etc., ibid. n° 19 bis.

78. We will now consider briefly the corresponding questions

for number-fields; for these, unfortunately, conjecture (cf. J. Milton, Paradise Regained, IV. 292) must replace knowledge at the crucial point.

In order to discuss functional equations in §75, we assumed that k was a function-field; for a number-field, the infinite places and their "gamma factors" must be taken into account; we first recall their definition for "ordinary" L-series. For any quasicharacter α of $k_A^\times/k^\times$, we put

$$G_k(\alpha) = \prod_w [\pi \mathcal{G}_{k_w}(\alpha_w)] ,$$

where the product is taken over the infinite places of k; for $k_w = \mathbf{R}$ resp. $\mathbf{C}$, we take here for $\mathcal{G}_{\mathbf{R}}$ resp. $\mathcal{G}_{\mathbf{C}}$ the function defined in §46, a). We will write $G_k(\alpha, s)$ for $G_k(\alpha, \omega_s)$.

Now (following T. Tamagawa, J. Fac. Sc. Tokyo 6 (1953), pp. 421-428) we attach to each finite extension k' of k, and to each character χ of $\mathfrak{G}_{k'}$, a meromorphic function $s \longrightarrow G_{k'}(\chi', s)$, so as to fulfil the following conditions, similar to those in §73: (I) for a prime character α' of degree 1 of $\mathfrak{G}_{k'}$, $G_{k'}(\alpha', s)$ is as defined above; (II) $\chi' \longrightarrow G_{k'}(\chi', s)$ is a morphism of $X_{k'}$ into the multiplicative group of meromorphic functions of s; (III) $G_{k'}([\chi''; k''/k'], s)$ is the same as $G_{k''}(\chi'', s)$. These conditions are compatible (because of the identity $\pi G_2(s) = \pi G_1(s). \pi G_1(s+1)$, and because we inserted the factors π into the definition of $G_k(\alpha)$). More precisely, if χ is the character of a representation M of $\mathfrak{G}_{k'}$, $G_k(\chi, s)$ is a product of factors $G_w(\chi, s)$ for the infinite places w of k, each of which depends only upon the restriction of M to the group $\mathfrak{G}_{k,w}$ introduced in §73. For $k_w = \mathbf{C}$, $\mathfrak{G}_{k,w}$ is isomorphic to $\mathbf{C}^\times$; then, if χ is of degree n, it decomposes on $\mathfrak{G}_{k,w}$ into a sum of n quasicharacters α_i of $\mathbf{C}^\times$, and $G_w(\chi, s)$ is the product $\prod [\pi \mathcal{G}_{\mathbf{C}}(\alpha_i \omega_s)]$. If $k_w = \mathbf{R}$, $\mathfrak{G}_{k,w}$ is the semidirect product of $\mathbf{C}^\times$ with a group $\{1, e\}$, defined by the relations

$e^2 = -1$, $e^{-1}xe = \bar{x}$ for $x \in \mathbb{C}^\times$; this has a morphism ν onto $\mathbb{R}^\times$, given by $\nu(e) = -1$, $\nu(x) = x\bar{x}$, and its irreducible representations are either of degree 2 and of the form $[\alpha'; \mathbb{C}/\mathbb{R}]$, where α' is a quasicharacter of $\mathbb{C}^\times$ such that $\alpha' \neq \bar{\alpha}'$, or of degree 1 and of the form $\beta \circ \nu$, where β is a quasicharacter of $\mathbb{R}^\times$; to these, we attach respectively the factors $\pi\,\mathcal{G}_{\mathbb{C}}(\alpha'\omega_s)$ and $\pi\,\mathcal{G}_{\mathbb{R}}(\beta\omega_s)$, and define $G_w(\chi, s)$ accordingly. Thus, in all cases, $G_k(\chi, s)$ is a product of functions of the form $\Gamma(s+a)$ and $\Gamma((s+b)/2)$ and of an exponential e^{cs+d}, and its reciprocal is an entire function.

Now we put:

$$\Lambda(\chi, s) = L(\chi, s)G_k(\chi, s), \quad \Lambda_o(\chi) = \Lambda(\chi, 1/2),$$

where, as before, $L(\chi, s)$ is the same as $L_k(\chi\omega_s)$. We can repeat lemma A and its proof, merely replacing L_o by Λ_o, and applying the functional equation for the "ordinary" L-series. This defines a morphism η of X_k into $\mathbb{C}^\times$.

79. Define $H(\chi, \chi')$ as in the proof of lemma B. One finds immediately that, if α, β are two quasicharacters of $k_A^\times/k^\times$ with disjoint conductors, $H(\alpha, \beta)$ is always ± 1, but not necessarily 1. We observe, however, that the proof of lemma B can be applied word for word to $H(\chi, \chi')^2$, except that for this the last argument under d), concerning $H(k', k'')$, becomes superfluous since $H(k', k'')$ is trivially ± 1. Consequently, we have at any rate $H(\chi, \chi') = (-1)^{B(\chi, \chi')}$ where B is a symmetric bilinear mapping of $X_k \times X_k$ into the field $\mathbb{F}_2$ with two elements.

Now one verifies easily the following facts:

a) If α, β are quasicharacters of $k_A^\times/k^\times$, then $B(\alpha, \beta)$ can be written as a sum $\Sigma B_w(\alpha_w, \beta_w)$ over the infinite places w of k, where the terms are as follows: if $k_w = \mathbb{R}$, $B_w(\alpha_w, \beta_w)$ is 1 if

$\alpha_w(-1) = \beta_w(-1) = -1$, and 0 otherwise; if $k_w = \mathbf{C}$ and α_w, β_w are written in the form $x^M(x\bar{x})^s$, $x^N(x\bar{x})^t$ respectively, with M, N in $\mathbf{Z}$ and s, t in $\mathbf{C}$, then $B_w(\alpha_w, \beta_w)$ is 0 if $MN \geqq 0$ and $\equiv \inf(|M|, |N|) \bmod. 2$ if $MN < 0$.

b) If α is a quasicharacter of $k_A^\times/k^\times$ and k' a finite extension of k, then $B(\alpha, [1; k'/k])$ can be written as a sum $\Sigma B_w(\alpha_w; k')$, where the terms are as follows: if $k_w = \mathbf{R}$, $\alpha_w(-1) = -1$, and N is the number of places w' of k' above w such that $k'_{w'} = \mathbf{C}$, then $B_w(\alpha_w, k')$ is $\equiv N \bmod. 2$; otherwise it is 0.

c) Formula (53) of §75, b) is valid without any change, so that B satisfies a similar formula:

$$B(\chi, [\beta'; k'/k]) = B_{k'}(\chi', \beta') + B(\chi, [1; k'/k]) .$$

d) Using these facts, and using Brauer's theorem just as in §75, d), one finds now that $B(\chi, \chi')$ is a sum $\Sigma B_w(\chi, \chi')$ taken over the infinite places w of k, where $B_w(\chi, \chi')$ is a symmetric bilinear form on $X_k \times X_k$, depending only upon the restrictions of χ, χ' to $\mathfrak{G}_{k,w}$. Thus, in order to determine $B(\chi, \chi')$ completely, it is enough to indicate the values of B_w for a pair of irreducible characters of $\mathfrak{G}_{k,w}$. For $k_w = \mathbf{C}$, such characters are all of degree 1; then B_w is as stated above under a). For $k_w = \mathbf{R}$, we have to deal (as explained above) with characters of the form $[\alpha'; \mathbf{C}/\mathbf{R}]$ or $\beta \circ \nu$; then, using a), b) and c) above, one finds at once the following:
$B_w(\beta \circ \nu, \beta' \circ \nu)$ is 1 if $\beta(-1) = \beta'(-1) = -1$, and 0 otherwise;
$B_w(\beta \circ \nu, [\alpha'; \mathbf{C}/\mathbf{R}])$ is 1 if $\beta(-1) = -1$, and 0 otherwise;
$B_w([\alpha'; \mathbf{C}/\mathbf{R}], [\alpha''; \mathbf{C}/\mathbf{R}])$ is the same as $B_w(\alpha', \alpha'')$ when $k_w = \mathbf{C}$.
These results take the place of lemma B.

80. Just as in §76, we consider now a positive character χ of degree 2 of $\mathfrak{G}_k$ and define two Dirichlet series Z, Z' by

$$Z(s) = L_o(\chi\omega_s) , \quad Z'(s) = \eta'(\chi)L_o(\hat{\chi}\omega_s) ,$$

with a constant $\eta'(\chi)$ which will be defined presently; the corresponding extended Dirichlet series are then $L_o(\chi\omega)$ and $\eta'(\chi)L_o(\hat{\chi}\omega)$. As in §76, these series are eulerian at all finite places of k outside $\mathfrak{f}_\chi$; the Euler factors have the form prescribed by §§24-25, with $a = \delta_\chi$, and λ as in §76.

We have seen in §74 that $Z(s)$, $Z'(s)$ can be continued to the whole plane, and $Z(\omega)$, $Z'(\omega)$ to the whole group Ω_k, as meromorphic functions; in §78 it has been shown that, when they are multiplied with the proper gamma factors, they satisfy functional equations of the familiar type, and §79 supplies us with information concerning the constant factors in these equations. We may now seek to match these gamma factors, and these constant factors, with those occurring in theorems 6 and 7 of Chapter X. After some trivial calculations, one does find that Z, Z' satisfy the functional equation in these theorems for a type of automorphic pairs defined as follows. As before, we take $a = \delta_\chi$ and $\mathfrak{A} = \mathfrak{f}_\chi$. As to the choice of types at the infinite places, this is the recipe:

a) For $k_w = \mathbf{R}$, the restriction of χ to $\mathfrak{G}_{k,w}$ must either be of the form $[\alpha'; \mathbf{C}/\mathbf{R}]$ or must split into two prime characters α_1, α_2 of degree 1; in the former case, if $\alpha'(x) = (x\bar{x})^s$, χ splits into $\alpha_1(x) = |x|^s$ and $\alpha_2(x) = (\mathrm{sgn}\ x).|x|^s$; therefore, in the latter case, we may assume that $\alpha_1\alpha_2^{-1}$ is not $x \longrightarrow \mathrm{sgn}\ x$. That being so, in the former case, put $\alpha'(x) = x^{\zeta'}\bar{x}^{\zeta''}$; then put $\rho = |\zeta' - \zeta''|$, $\delta = \frac{1}{2}(\rho^2 - 1)$, $n = \rho + 1$, and associate with that place the discrete type (δ_χ, δ, n). Also, put $e_w = i^{-1-\rho}$. In the latter case, write $\alpha_1(x)^{-1}\alpha_2(x) = (\mathrm{sgn}\ x)^m |x|^\rho$, where we assume (after interchanging α_1 and α_2 if necessary) that $\mathrm{Re}(\rho) \geqq 0$. Put $\delta = \frac{1}{2}(\rho^2 - 1)$, $\alpha_2(-1) = (-1)^d$, and associate with w the principal type $(\delta_\chi, \delta, d, m)$. Also, put $e_w = \kappa_w(\alpha_1)\kappa_w(\alpha_2)$.

b) For $k_w = \mathbb{C}$, the restriction of χ to $\mathfrak{g}_{k,w}$ splits into two prime characters α_1, α_2. Put $\alpha_1(x)^{-1}\alpha_2(x) = x^{\rho'}\bar{x}^{\rho''}$; as before, we may assume $\mathrm{Re}(\rho' + \rho'') \geqq 0$. Then put $\delta' = \frac{1}{2}(\rho'^2 - 1)$, $\delta'' = \frac{1}{2}(\rho''^2 - 1)$, $n = |\rho' - \rho''|$; associate with the place w the type $(\delta_\chi, \delta', \delta'', n)$ and put $e_w = \kappa_w(\alpha_1)\kappa_w(\alpha_2)$.

Finally, take $\eta'(\chi) = \eta(\chi)\prod e_w$ in the definition of Z'.

As pointed out in Remark 1, §49, we could, in each one of the above cases, change the choice of n, subject to the conditions indicated there, without affecting the functional equation.

81. All necessary conditions for applying theorem 7 to Z and Z' have now been verified, except the decisive one: both sides of the functional equation should be holomorphic (except for at most finitely many poles) and bounded in every strip. Using an argument of the Phragmèn-Lindelöf type (similar to lemma 11 of §66) one sees that the latter condition would follow from the former one, since it is satisfied by the "ordinary" L-functions. What remains is to verify Artin's conjecture, at any rate for characters χ of degree 2. Until that is done, the relation between the series Z, Z' and automorphic forms can only be stated when $L(\chi)$ is either a product of two L-functions, i.e. when χ splits into two characters α_1, α_2 of degree 1, or when it is an L-function over a quadratic extension k' of k, i.e. when χ is of the form $[\alpha'; k'/k]$. In the former case (just as in §77) one must assume, in order to apply theorem 7, that α_1 and α_2 are not unramified; if they were so, one would need the converse to the results of §65 in Chapter X; here again, Jacquet and Langlands have shown that the automorphic forms Φ, Φ' are Eisenstein series. If $\chi = [\alpha'; k'/k]$ for a quadratic extension of k, theorem 7 can always be applied.

As observed in §70, all zeta-functions of elliptic curves with complex multiplication belong to one of the above types, with

$\chi = [\mathfrak{a}'; k'/k]$ or $\chi = \mathfrak{a}_1 + \mathfrak{a}_2$ according as the compositum k' of k and of the imaginary quadratic field of complex multiplications is of degree 2 or 1 over k.

Bitte wenden / Continued

Vol. 72: The Syntax and Semantics of Infinitary Languages. Edited by J. Barwise. IV, 268 pages. 1968. DM 18,– / $ 5.00

Vol. 73: P. E. Conner, Lectures on the Action of a Finite Group. IV, 123 pages. 1968. DM 10,– / $ 2.80

Vol. 74: A. Fröhlich, Formal Groups. IV, 140 pages. 1968. DM 12,– / $ 3.30

Vol. 75: G. Lumer, Algèbres de fonctions et espaces de Hardy. VI, 80 pages. 1968. DM 8,– / $ 2.20

Vol. 76: R. G. Swan, Algebraic K-Theory. IV, 262 pages. 1968. DM 18,– / $ 5.00

Vol. 77: P.-A. Meyer, Processus de Markov: la frontière de Martin. IV, 123 pages. 1968. DM 10,– / $ 2.80

Vol. 78: H. Herrlich, Topologische Reflexionen und Coreflexionen. XVI, 166 Seiten. 1968. DM 12,– / $ 3.30

Vol. 79: A. Grothendieck, Catégories Cofibrées Additives et Complexe Cotangent Relatif. IV, 167 pages. 1968. DM 12,– / $ 3.30

Vol. 80: Seminar on Triples and Categorical Homology Theory. Edited by B. Eckmann. IV, 398 pages. 1969. DM 20,– / $ 5.50

Vol. 81: J.-P. Eckmann et M. Guenin, Méthodes Algébriques en Mécanique Statistique. VI, 131 pages. 1969. DM 12,– / $ 3.30

Vol. 82: J. Wloka, Grundräume und verallgemeinerte Funktionen. VIII, 131 Seiten. 1969. DM 12,– / $ 3.30

Vol. 83: O. Zariski, An Introduction to the Theory of Algebraic Surfaces. IV, 100 pages. 1969. DM 8,– / $ 2.20

Vol. 84: H. Lüneburg, Transitive Erweiterungen endlicher Permutationsgruppen. IV, 119 Seiten. 1969. DM 10,– / $ 2.80

Vol. 85: P. Cartier et D. Foata, Problèmes combinatoires de commutation et réarrangements. IV, 88 pages. 1969. DM 8,– / $ 2.20

Vol. 86: Category Theory, Homology Theory and their Applications I. Edited by P. Hilton. VI, 216 pages. 1969. DM 16,– / $ 4.40

Vol. 87: M. Tierney, Categorical Constructions in Stable Homotopy Theory. IV, 65 pages. 1969. DM 6,– / $ 1.70

Vol. 88: Séminaire de Probabilités III. IV, 229 pages. 1969. DM 18,– / $ 5.00

Vol. 89: Probability and Information Theory. Edited by M. Behara, K. Krickeberg and J. Wolfowitz. IV, 256 pages. 1969. DM 18,– / $ 5.00

Vol. 90: N. P. Bhatia and O. Hajek, Local Semi-Dynamical Systems. II, 157 pages. 1969. DM 14,– / $ 3.90

Vol. 91: N. N. Janenko, Die Zwischenschrittmethode zur Lösung mehrdimensionaler Probleme der mathematischen Physik. VIII, 194 Seiten. 1969. DM 16,80 / $ 4.70

Vol. 92: Category Theory, Homology Theory and their Applications II. Edited by P. Hilton. V, 308 pages. 1969. DM 20,– / $ 5.50

Vol. 93: K. R. Parthasarathy, Multipliers on Locally Compact Groups. III, 54 pages. 1969. DM 5,60 / $ 1.60

Vol. 94: M. Machover and J. Hirschfeld, Lectures on Non-Standard Analysis. VI, 79 pages. 1969. DM 6,– / $ 1.70

Vol. 95: A. S. Troelstra, Principles of Intuitionism. II, 111 pages. 1969. DM 10,– / $ 2.80

Vol. 96: H.-B. Brinkmann und D. Puppe, Abelsche und exakte Kategorien, Korrespondenzen. V, 141 Seiten. 1969. DM 10,– / $ 2.80

Vol. 97: S. O. Chase and M. E. Sweedler, Hopf Algebras and Galois theory. II, 133 pages. 1969. DM 10,– / $ 2.80

Vol. 98: M. Heins, Hardy Classes on Riemann Surfaces. III, 106 pages. 1969. DM 10,– / $ 2.80

Vol. 99: Category Theory, Homology Theory and their Applications III. Edited by P. Hilton. IV, 489 pages. 1969. DM 24,– / $ 6.60

Vol. 100: M. Artin and B. Mazur, Etale Homotopy. II, 196 Seiten. 1969. DM 12,– / $ 3.30

Vol. 101: G. P. Szegö et G. Treccani, Semigruppi di Trasformazioni Multivoche. VI, 177 pages. 1969. DM 14,– / $ 3.90

Vol. 102: F. Stummel, Rand- und Eigenwertaufgaben in Sobolewschen Räumen. VIII, 386 Seiten. 1969. DM 20,– / $ 5.50

Vol. 103: Lectures in Modern Analysis and Applications I. Edited by C. T. Taam. VII, 162 pages. 1969. DM 12,– / $ 3.30

Vol. 104: G. H. Pimbley, Jr., Eigenfunction Branches of Nonlinear Operators and their Bifurcations. II, 128 pages. 1969. DM 10,– / $ 2.80

Vol. 105: R. Larsen, The Multiplier Problem. VII, 284 pages. 1969. DM 18,– / $ 5.00

Vol. 106: Reports of the Midwest Category Seminar III. Edited by S. Mac Lane. III, 247 pages. 1969. DM 16,– / $ 4.40

Vol. 107: A. Peyerimhoff, Lectures on Summability. III, 111 pages. 1969. DM 8,– / $ 2.20

Vol. 108: Algebraic K-Theory and its Geometric Applications. Edited by R. M. F. Moss and C. B. Thomas. IV, 86 pages. 1969. DM 6,– / $ 1.70

Vol. 109: Conference on the Numerical Solution of Differential Equations. Edited by J. Ll. Morris. VI, 275 pages. 1969. DM 18,– / $ 5.00

Vol. 110: The Many Facets of Graph Theory. Edited by G. Chartrand and S. F. Kapoor. VIII, 290 pages. 1969. DM 18,– / $ 5.00

Vol. 111: K. H. Mayer, Relationen zwischen charakteristischen Zahlen. III, 99 Seiten. 1969. DM 8,– / $ 2.20

Vol. 112: Colloquium on Methods of Optimization. Edited by N. N. Moiseev. IV, 293 pages. 1970. DM 18,– / $ 5.00

Vol. 113: R. Wille, Kongruenzklassengeometrien. III, 99 Seiten. 1970. DM 8,– / $ 2.20

Vol. 114: H. Jacquet and R. P. Langlands, Automorphic Forms on GL (2). VII, 548 pages. 1970. DM 24,– / $ 6.60

Vol. 115: K. H. Roggenkamp and V. Huber-Dyson, Lattices over Orders I. XIX, 290 pages. 1970. DM 18,– / $ 5.00

Vol. 116: Séminaire Pierre Lelong (Analyse) Année 1969. IV, 195 pages. 1970. DM 14,– / $ 3.90

Vol. 117: Y. Meyer, Nombres de Pisot, Nombres de Salem et Analyse Harmonique. 63 pages. 1970. DM 6,– / $ 1.70

Vol. 118: Proceedings of the 15th Scandinavian Congress, Oslo 1968. Edited by K. E. Aubert and W. Ljunggren. IV, 162 pages. 1970. DM 12,– / $ 3.30

Vol. 119: M. Raynaud, Faisceaux amples sur les schémas en groupes et les espaces homogènes. III, 219 pages. 1970. DM 14,– / $ 3.90

Vol. 120: D. Siefkes, Büchi's Monadic Second Order Successor Arithmetic. XII, 130 Seiten. 1970. DM 12,– / $ 3.30

Vol. 121: H. S. Bear, Lectures on Gleason Parts. III, 47 pages. 1970. DM 6,– / $ 1.70

Vol. 122: H. Zieschang, E. Vogt und H.-D. Coldewey, Flächen und ebene diskontinuierliche Gruppen. VIII, 203 Seiten. 1970. DM 16,– / $ 4.40

Vol. 123: A. V. Jategaonkar, Left Principal Ideal Rings. VI, 145 pages. 1970. DM 12,– / $ 3.30

Vol. 124: Séminare de Probabilités IV. Edited by P. A. Meyer. IV, 282 pages. 1970. DM 20,– / $ 5.50

Vol. 125: Symposium on Automatic Demonstration. V, 310 pages. 1970. DM 20,– / $ 5.50

Vol. 126: P. Schapira, Théorie des Hyperfonctions. XI, 157 pages. 1970. DM 14,– / $ 3.90

Vol. 127: I. Stewart, Lie Algebras. IV, 97 pages. 1970. DM 10,– / $ 2.80

Vol. 128: M. Takesaki, Tomita's Theory of Modular Hilbert Algebras and its Applications. II, 123 pages. 1970. DM 10,– / $ 2.80

Vol. 129: K. H. Hofmann, The Duality of Compact Semigroups and C*-Bigebras. XII, 142 pages. 1970. DM 14,– / $ 3.90

Vol. 130: F. Lorenz, Quadratische Formen über Körpern. II, 77 Seiten. 1970. DM 8,– / $ 2.20

Vol. 131: A Borel et al., Seminar on Algebraic Groups and Related Finite Groups. VII, 321 pages. 1970. DM 22,– / $ 6.10

Vol. 132: Symposium on Optimization. III, 348 pages. 1970. DM 22,– / $ 6.10

Vol. 133: F. Topsøe, Topology and Measure. XIV, 79 pages. 1970. DM 8,– / $ 2.20

Vol. 134: L. Smith, Lectures on the Eilenberg-Moore Spectral Sequence. VII, 142 pages. 1970. DM 14,– / $ 3.90

Vol. 135: W. Stoll, Value Distribution of Holomorphic Maps into Compact Complex Manifolds. II, 267 pages. 1970. DM 18,– / $

Vol. 136: M. Karoubi et al., Séminaire Heidelberg-Saarbrücken-Strasbourg sur la K-Théorie. IV, 264 pages. 1970. DM 18,– / $ 5.00

Vol. 137: Reports of the Midwest Category Seminar IV. Edited by S. MacLane. III, 139 pages. 1970. DM 12,– / $ 3.30

Vol. 138: D. Foata et M. Schützenberger, Théorie Géométrique des Polynômes Eulériens. V, 94 pages. 1970. DM 10,– / $ 2.80

Vol. 139: A. Badrikian, Séminaire sur les Fonctions Aléatoires Linéaires et les Mesures Cylindriques. VII, 221 pages. 1970. DM 18,– / $ 5.00

Vol. 140: Lectures in Modern Analysis and Applications II. Edited by C. T. Taam. VI, 119 pages. 1970. DM 10,– / $ 2.80

Vol. 141: G. Jameson, Ordered Linear Spaces. XV, 194 pages. 1970. DM 16,– / $ 4.40

Vol. 142: K. W. Roggenkamp, Lattices over Orders II. V, 388 pages. 1970. DM 22,– / $ 6.10

Vol. 143: K. W. Gruenberg, Cohomological Topics in Group Theory. XIV, 275 pages. 1970. DM 20,– / $ 5.50

Lecture Notes in Physics

Bisher erschienen/Already published

Vol. 1: J. C. Erdmann, Wärmeleitung in Kristallen, theoretische Grundlagen und fortgeschrittene experimentelle Methoden. II, 283 Seiten. 1969. DM 20,– / $ 5.50

Vol. 2: K. Hepp, Théorie de la renormalisation. III, 215 pages. 1969. DM 18,– / $ 5.00

Vol. 3: A. Martin, Scattering Theory: Unitarity, Analytic and Crossing. IV, 125 pages. 1969. DM 14,– / $ 3.90

Vol. 4: G. Ludwig, Deutung des Begriffs physikalische Theorie und axiomatische Grundlegung der Hilbertraumstruktur der Quantenmechanik durch Hauptsätze des Messens. XI, 469 Seiten. 1970. DM 28,– / $ 7.70

Vol. 5: M. Schaaf, The Reduction of the Product of Two Irreducible Unitary Representations of the Proper Orthochronous Quantummechanical Poincaré Group. IV, 120 pages. 1970. DM 14,– / $ 3.90

Vol. 6: Group Representations in Mathematics and Physics. Edited by V. Bargmann. V, 340 pages. 1970. DM 24,– / $ 6.60